CAMBRIDGE SCIEN(

Series editor Richard Ingle

Practical \
Sc

Brian Woolnough and Terry Allsop

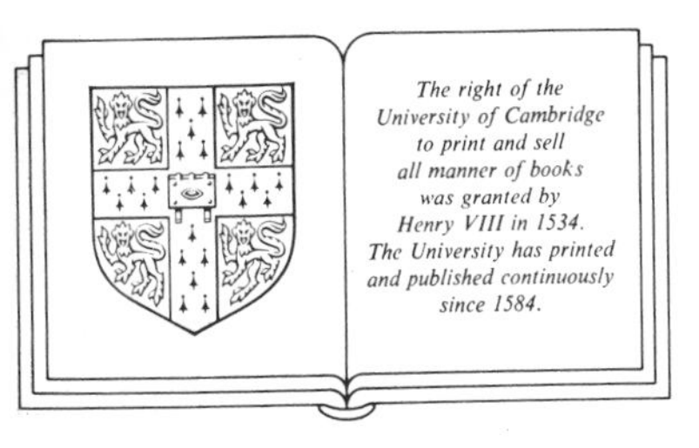

CAMBRIDGE UNIVERSITY PRESS

Cambridge
London New York New Rochelle
Melbourne Sydney

Published by the Press Syndicate of the University of Cambridge
The Pitt Building, Trumpington Street, Cambridge CB2 1RP
32 East 57th Street, New York, NY 10022, USA
10 Stamford Road, Oakleigh, Melbourne 3166, Australia

First published 1985

Printed in Great Britain at the University Press, Cambridge

Library of Congress catalogue card number: 84–28577

British Library cataloguing in publication data

Woolnough, Brian
Practical work in science. — (Cambridge science education series)
1. Science — Study and teaching (Secondary) — Great Britain
I. Title II. Allsop, Terry
507′.1241 Q183.4.G7

ISBN 0 521 27861 9

DS

CONTENTS

The authors

Brian Woolnough is a University Lecturer in Science Education in the Department of Educational Studies, Oxford University. After graduating from Reading University, and taking his PGCE at Cambridge, he taught physics and science in secondary schools in the UK. His current interests lie in the analysis of school science teaching, the supply and training of science teachers, and the place of practical work and technology in schools. He is particularly interested in the development of physics teaching in schools both nationally and internationally.

Terry Allsop is also a University Lecturer in Science Education at Oxford University. He graduated in Natural Sciences at Cambridge University. After teaching in England and involvement with the development of Nuffield Chemistry, he held positions in teacher education in Uganda and Hong Kong. He is interested in many aspects of science teacher education and science curriculum development, particularly in the place of laboratory work, in integrated science and in novel approaches to in-service education.

The series editor

Dr Richard Ingle is Lecturer in Science Education at the University of London Institute of Education. He graduated in the physical sciences at Durham University and then taught science in secondary schools for a period of fourteen years in Scotland, England and Uganda. He subsequently held posts in chemical education at Makerere University College, Uganda, and at the Centre for Science Education, Chelsea College, University of London. During the 1970s he undertook an evaluation of Nuffield Chemistry and subsequently became general editor of the revised Nuffield Chemistry series. He was for a time education adviser at the Ministry of Overseas Development. His current interests include the pre-service and in-service education of science teachers, cultural aspects of science education, and the probing of learning difficulties faced by pupils in using mathematics in the course of their science education.

ACKNOWLEDGEMENTS

The authors and the publisher would like to thank the following for permission to reproduce copyright material:
Fig 1.1, McGraw-Hill Book Company (UK) Ltd; Fig 4.1 from *Science in Schools: Age 15 No. 2*, Department of Education and Science 1984, reproduced with the permission of the Controller of Her Majesty's Stationery Office; Fig 5.1, Hodder and Stoughton Educational; Fig 5.2, Macdonald and Company (Publishers) Ltd.

The cover photograph shows girls at Stantonbury Campus, Milton Keynes, engaged in investigational project work.
Photograph by Richard Ingle.
Cover design by Andrew Bonnett.

PREFACE

Sine experimentia nihil sufficienter sciri potest

– without experiments nothing can be adequately known[1]

There is currently an active, lively and constructive debate concerning the nature and style of science teaching most appropriate to the students in our schools and colleges. Should science concentrate on content or process? Should we provide an education *in* science or use our subject to provide education *through* science? Should our science remain academic or become more technological? Should its aims be educational or vocational? How and when should students consider the implications of science in society? Such questions have been with science education throughout its short history, but with science now demanded for all pupils up to the age of sixteen, including the full primary age range, the questions now become more acute. At the centre of the debate about the nature of science teaching must be the role of practical work, for – in the United Kingdom particularly – a very large proportion of the time and effort devoted to science teaching is spent on small-group practical work. Science is a practical activity which takes place in a laboratory. As Solomon has said, 'Science simply belongs there as naturally as cooking belongs in a kitchen and gardening in a garden'.[2] But while few doubt that practical work has a place in the teaching of science, we believe that there is a very real need to rethink the most appropriate purposes for practical work and the forms which it should take. We believe that there is at present much confusion about, and indiscriminate use of, class practical work in science teaching, which leads to inefficient management of time and resources.

We believe that now is the time to re-evaluate that experience, in order to decide what type of practical is, and what is not, suitable for different purposes. In this book we shall be sharing some of our concerns about current practice and shall be indicating a more constructive approach to practical work.

It is always difficult to give acknowledgement to all the many people who have influenced our thinking over the years. But we wish to thank in particular Edward Black, Ros Driver, Lewis Elton,

Elizabeth Hitchfield, Richard Ingle and Jon Ogborn, who have pro vided most helpful comments on parts, or drafts, of this manuscript To the others, colleagues and friends, who have helped and stimu lated our thinking, either consciously or unconsciously, we acknow ledge our gratitude. The responsibility for the final product, o course, is ours – we hope it may help to stimulate further discussion

B.E.W
October 1984 R.T.A

Notes and References

1 Engraved over the doorway of the Daubeny Building, Oxford. This was used fo some of the earliest science teaching in the University.

2 J. Solomon *Teaching Children in the Laboratory*, Croom Helm (London) 198 page 13.

Problems and Possibilities in Current Practice

'wo science lessons seen on the same day in neighbouring schools pitomise something of the wide variety of practical work being done n schools. Both were in similar, well-ordered, happy classes of 12–3 year old students. Both were directed by experienced teachers vho were basing their work on a popular structured course. In the irst lesson the students were investigating invertebrates. They were ollecting insects from simple insect traps that they had previously et up in the school grounds. They were handling, cleaning, studying nd classifying their catch, and working with some worms from their mall wormery. In general they were acting like independent scienists, studying their environment. They were observing, recording, lassifying and communicating. They were handling animals and lanning, developing, and executing their own experiments with imple home-made apparatus. They were acquiring the processes nd skills of science and at the same time learning about and gaining onfidence in exploring their environment. The second class was folowing instructions from a worksheet which told them to use a unsen burner to heat different amounts of water in beakers for the ame length of time and to record the temperatures before and after eating. By following the operation carefully step by step, the teacher oped that the students would come to understand the difference etween heat and temperature. It was clear by the end of the lesson hat the students had not appreciated this subtle distinction, indeed hey had not even realised why they had been heating the water in the eakers. They were certainly not acting like problem-solving scienists, nor had they learnt any physics!

Current Practice

n the United Kingdom, perhaps more than in other European counries, a great amount of time and effort is devoted to small-group pracical work in the teaching of science. It is not uncommon for more

than a third of the time of 16–18 year old scientists to be spent o
practical work,[1] while most 11–13 year olds will spend well ove
half of their science lessons doing practical.[2] Children in primar
schools are becoming increasingly involved with scientific activity
while science students in higher education have traditionally foun
the practical class an essential, if low status and largely unexplained
part of their courses. As a high proportion of teaching time is spent o
practical work, so is an even higher proportion of cost in terms o
money and in effort. There is a clear commitment by science teacher
to the value of practical as a central part of a scientific education. An
yet we would want to question whether, in practice, all of that time
money and effort is well spent. We would want to suggest that w
need to rethink the purposes of practical work in science, and then t
consider the most appropriate kind of practical which will fulfi
those aims. At present most practical work is fitted into the content
led curriculum and used to illustrate aspects of scientific knowledg
and theory. Much practical work appears to pupils to be a successio
of exercises, with apparatus, through which they are led in the hop
of solving an unasked question. Her Majesty's Inspectors of Schools
in their survey of secondary schools in 1977, reported that

> Many science teachers recognised the importance of practical work. They be
> lieved that pupils should have first-hand practical experience in laboratories i
> order to acquire skills in handling apparatus, to measure and to illustrate con
> cepts and principles. Unfortunately practical work often did not go further tha
> this and few opportunities were provided for pupils to conduct challengin
> investigations.[3]

Although this is a comment on the schools of England and Wales, i
could well apply to practice in many other countries. For example,
recent analysis of science education in Canadian schools reports tha

> work in the lab is geared towards illustrating facts and theories presented in th
> class room, confirming what is discussed in class, obtaining precise facts and get
> ting the right answers to problems . . . , teachers emphasise routines, standard
> of accuracy and thoroughness . . . This emphasis on approved explanations an
> the right answer is at odds with the process of inquiry and the conceptual an
> tentative status of knowledge in science. Yet, such predictable activities as note
> taking, copying activity sheets and lab procedures are valued because th
> accumulated information provides a base for work in the next grade, and becaus
> they control and channel energies by keeping students busy with routine an
> unambiguous work.[4]

We believe that such descriptions summarise current practic
throughout much of the world. At a time when practical work is com
ing under increasing pressure because of its very heavy demands o

inancial resources in a time of economic constraint, it is, increasngly, important to be clear that we have sound reasons for doing it.

Underlying Concerns

Underlying our concern about much of the practical work in schools today we see three major issues.

First, much practical work in schools appears to have very little to do with the activity of practising scientists; it is much more a series of restrictive exercises involving students in activities in a science laboratory, with school science apparatus, on a range of scientific topics. It is often closed, convergent and dull. The question is how can it be made divergent and exhilarating? It is a travesty of what science should be about and leaves little but memories of trivia in the minds of many students who have experienced it. Previous generations spoke critically of experiments with optical pins, copper calorimeters, burettes and pipettes. Current students speak of incomprehensible exercises with ticker timers and ripple tanks. Such practical work sees science as a means of acquiring knowledge, the process of science as leading inexorably to right knowledge. Though much activity is involved, this activity is designed to lead to the 'correct' answer, the means being subservient to the end. And underlying it all is the philosophy that 'knowing is more important than doing'. It assumes that scientific knowledge is objective and detached, and learning science a matter of attaching that which is known by others onto students who previously did not know it. Can we not make practical work for our students more like 'real science'?

Secondly, and this most particularly refers to that large proportion of practical work which is designed to elucidate theory, there is a growing body of research evidence which suggests that current practices just do not work: teaching theory through practical work is not an efficient way of transmitting an understanding of scientific concepts to students, indeed it can be positively harmful. We know that many of our students come through a practically based, theoretically orientated, course with only a very modest understanding of scientific theory. Summarising the work of those who have researched into students' understanding of scientific ideas, Osborne and Wittrock conclude that young children have firmly held views about many science topics prior to being taught science, that their ideas can be amazingly tenacious and resistant to change, and that, in some cases, older children have ideas which are *less* in agreement with the views of scientists than the views of younger children.[5] Any study of a student's understanding of, say, basic electricity after spending weeks working with circuit boards, or of acceleration after experimenting with ticker timers and trolleys, will soon shatter the illusion that small-group practical work is an effective way of teaching theories in physics. For students to appreciate and master theoretical concepts, they need to handle them at the abstract level. Could it be that the distracting effect of the concrete can, at times, be positively misleading and restrictive?

Our third concern relates to the enormous amount of time, effort and money spent by teachers on practical work. The planning, preparing, and finding of ingenious ways of getting students through it, prevent us from giving adequate thought to alternative strategies for teaching, even for alternative approaches to practical work. Some teachers consider that all of their teaching should be through practical work, and even feel guilty if their students are not 'doing practical'. This has the effect of restricting what is taught in science lessons to those topics which can be studied practically by students in the laboratory. Currently there is concern that our science teaching as a whole is too narrow, that we should be widening our frontiers beyond 'science for the enquiring mind', to encompass 'science for action' (its applications in the world) and 'science for the citizen' (its implications for society).[6] The teaching strategies suited to these wider aims are quite different from the traditional small-group practical, and may involve discussions, debates, simulations, visits, decision-making games – a whole range of techniques for engaging the pupils' minds which is used, with profit, in other areas of the school curriculum. While so many science lessons are dominated by practical work, teachers have little time or inclination to explore ways of incorporating a wider variety of teaching strategies to match the wider aims.

The cost of running a specialist science laboratory with most of the course work based on small-group practical is considerable. Indeed for most countries it is becoming prohibitive. So the question must be asked, 'Do we need laboratories?'. Richmond has argued that the answer may be 'no'.[7] Although he was arguing largely in the context of the developing countries, his thesis that there is sufficient experience and apparatus around in the everyday world of students on which to build a base for those theories that are necessary for a scientific education may be equally valid in a wider context. Indeed, as Jennings argues in another book in this series, *Science in the Locality*,[8] there is enormous, untapped potential for valuable and interesting scientific activities within the locality of every school. Toothacker similarly asks the question 'Is student lab work worth this great expense of time and money?', answering 'No – at least for the kind of laboratory work currently done'.[9] Others would go further and state that the sophisticated apparatus found in most science laboratories restricts creativity and presents an additional stage in removing the science learned in schools from the student's real world. So the financial restrictions currently affecting the educational scene may yet bring us benefit – as Rutherford once said: 'We haven't the money so we will have to think!'

These then are some of the underlying concerns which encourage us to look more critically at the practice of practical work in schools. There is indeed much evidence of excellent practical work going on which is worth every penny spent on it, but we will return to that later. It is, of course, much too simple to talk of practical work as if it were all of the same kind, so we shall next analyse different types and consider their relative strengths and weaknesses.

Classification of Practical Work

When we look at the aims that teachers and curriculum developers give for doing small-group practical work, we find that they may fall under one of four main headings: aims concerned with motivational factors; aims concerned with developing experimental skills and techniques; aims concerned with the learning of the scientific approach ('being a scientist for a day'); and aims concerned with gaining a better understanding of the theoretical aspects of the course. We will look at each of these groups and consider the extent to which each can be fulfilled through practical work.

Motivational Factors

It is often stated that practical work should be done to interest and motivate students in science lessons. Although such statements reflect an out-dated stimulus–response model of teaching science, with motivation seen as an extrinsic factor added on to the learning process, many teachers still speak, and courses are written, in these terms. Kerr and others have shown how highly teachers rate motivational factors.[10]

Students, from the beginning of secondary school through to higher education, similarly justify the use of practical work in science teaching on the grounds of interest and motivation. Indeed they come to science lessons with the expectation that they will normally be doing practical work, so that if the teacher produces other, non-practical strategies, they may react negatively.

Experimental Skills and Techniques

Much of a scientist's work involves 'doing experiments', and certain skills and techniques are required before this can be carried out successfully. Skills of observation and measurement need to be developed, and also the realisation that we often see only that for which we are looking. Techniques for the safe and systematic handling of apparatus need to be developed. The task may be as straightforward as learning to filter a solution or heat a test-tube, or a more complex operation such as using a cathode ray oscilloscope to measure a frequency or learning basic microbiological techniques. The appropriate handling of apparatus, and the acquiring of habits of accuracy, tenacity and honesty, are all parts of the scientist's armoury.

'Being a Scientist for a Day'

Many aims centre around the theme of pupils learning to act like a scientist, to acquire the scientific approach, and the belief that this can only be done by actually doing practical work the scientific way. The phrase 'being a scientist for a day' was much used in the early days of the Nuffield Courses. Less clear is what it actually meant to be a scientist; different people had quite different ideas as to the nature of scientific activity and the way a scientist works. The Baconian model of proceeding by induction, where a scientist collects and orders a mass of observational data from which in due course a pattern or a theory evolves is rarely promoted now but this meaning could be inferred from much of the practical work currently being undertaken. More respectably, the science teacher working on hypothesis testing could claim that Popper's ideas are the basis of this approach. Popper claimed that scientists start their investigations already holding a theory, a hypothesis, derived from a mixture of past experience and personal creativity and then set about testing this hypothesis. They throw the cloak of familiarity over something and see if it fits. Popper, of course, accepted that no hypothesis could be proved true but that testing should attempt to *dis*prove a hypothesis by falsification. In practice, however, this model appears rarely to be considered in school practical work, where the purpose usually seems to be to verify a theory or to decide between rival hypotheses. A third model of the scientist in action is as a discoverer knowledge, the approach used by Armstrong (see Chapter 2) in his lvocacy of the heuristic method, a view of science which fits naturly into the child-centred, discovery tradition of educators such as ousseau and Dewey. Bruner has supported this approach, stressing at the student, as also the research scientist, would be personally eking to discover knowledge and regularities of previously unregnised relations and similarities between ideas. Such discovery odels were very influential in much of the curriculum development in science in the 1960s, though again there has been a distortion f the ideal into the 'guided discovery' or 'stage-managed heurism' pe of practical, far removed from personal discovery. A fourth, uite different, model of the scientist perceives science as a craft ctivity. This view has been articulated most clearly by Polanyi[11] and ibsequently by Ravetz,[12] who says 'scientific work is necessarily a raft activity, depending on a personal knowledge of particular iings and a subtle judgement of their properties'. Through experince a scientist would build up a personal, tacit feel for the materials nd concepts with which he is working and will develop a sense of ie appropriateness of what should be done in tackling a particular roblem. The scientist must be 'an accomplished craftsman, having ndergone an apprenticeship, learning how to do things without lways) being able to appreciate why they work'. The investigation f problems will be achieved by a range of 'methods which are mainly iformal and tacit'. We see examples of this model of science in the roject work of science in primary schools and in Nuffield Advanced iology and Physics.

So, we have seen that even within the aim of 'being a scientist for a lay' there are a number of different interpretations of the nature of cientific activity whose genesis can be traced back to Bacon, Popper, rmstrong or Polanyi. The type of practical intended would depend n the view of what constituted 'the scientific method'.

)oing Practical to Support Theory

Jnderlying many aims for doing practical work lies the assumption hat it will be done to support the theoretical concepts which form he framework for the subject matter. It might be done 'to verify the heory', 'to discover the theory', 'to elucidate the theory', but always he theory and practical are interwoven – with the practical work often seen as being introduced for the sake of underpinning theoretical development. In some courses that practical work would only be ustified if it fitted into the natural development of the theoretical

argument, while others covertly hold to this principle in the way th
practical work is always made to fit into the content framework of th
course. Indeed Kerr, in defending himself from a possible accusatio
that he might be '. . . seeming to support a division between theor
and practice', said 'such a divorce is not implied; each must suppo
the other to form an integrated experience'.[13] Aims in this area hav
been justified as increasing both the remembering of the conter
covered and the understanding of the underlying theory. The catc
phrase, 'I hear and I forget, I see and I remember, I do and I unde
stand' has been widely quoted as a justification of practical wor
Clearly 'guided discovery' and 'stage-managed heurism' accord wit
this approach as the discoverer is directed to discovering the rig
fact or the right theory. We shall argue that this tight coupling of pra
tical and theory can have a detrimental effect both on the quality
practical work done and on the theoretical understandings gained b
the students.

Does it Work?

The question we must now ask of practical work is 'does it work?', i.
does it fulfil its aims? And we will consider this question in turn f
each of our four main groups of aims.

The first group is focussed on interest and motivational factors. W
all know the delight and excitement that a young student brings t
early science lessons. But we probably also recognise the way i
which that excitement fades as exploration of personal interests ar
replaced by experiments that interest the teacher. Given a choice i
science between a double lesson of practical and a double lesson c
theory, most would prefer the former – but that may be irrelevant an
may simply say more about the need for appropriate alternativ
teaching strategies for teaching theory than about practical work. W
also can see the student, so often it is the girls, for whom practica
work appears to hold no interest at all – the main concern is in th
neat and correct writing up of the experiment. How often have w
heard, 'Why do we have to do it for ourselves? Why can't you just tel
us?' For more able students the pedestrian pace enforced by pre
programmed practical work in order to deduce what may be seen a
blindingly obvious can be very frustrating. Ausubel touches on thi
when questioning the desirability of the discovery method:

> Is (it) a feasible technique for transmitting the substantive content of an intelle
> tual or scientific discipline to cognitively mature students who have alread
> mastered its rudiments and basic vocabulary?[14]

For such students too the motivational value of much practical work can be very limited.

Our second group of aims concerns the development of practical skills and techniques. It may appear obvious that students will become more competent at handling apparatus through using it. But is this so? Questions still arise as to how good students are at, say, observing and measuring, and whether competence in handling special school apparatus has any relevance to competence in handling different, more elaborate equipment outside school. More fundamentally, are the higher skills of planning, executing and interpreting experiments and investigations developed through school practical work? In their work in science, the Assessment of Performance Unit (A.P.U.) investigates performance in the practical skills of 'using apparatus and measuring instruments', 'using observations' and 'designing and performing investigations', skills fundamental to the practice of being a scientist. To date, results from their surveys[15] do not suggest that students have developed these skills fully by the ages of 13+ or 15+. Such findings as 'there seems to be a general absence of self-directed systematic observation', 'there was a tendency for pupils to give answers based on prior knowledge rather than observations made in the test situation', and 'generally low level of performance on subcategory 2B (estimating quantities)', led them to the conclusion that 'more attention might be given to levels of performance in basic practical skills involving more complex measurement techniques and skills of estimation. It should not be assumed that these will be acquired "in passing" in a science course'. Earlier work by Brown in the U.S.A., and Aspen with Open University students also showed, to quote Toothacker, that 'students who have laboratory classroom experience have not necessarily learned simple laboratory skills'.[16] Clearly we have no grounds for complacency. The A.P.U. have also stressed the need for more attention, and more opportunity, to be given to explicit teaching of the skills of experimental design. Her Majesty's Inspectors of Schools found little opportunity being given to students to devise their own experiments or to conduct challenging experimental investigations.[17] Most practical work made little demand on the pupils to do their own planning. Research by the Engineering Industry Training Board, who were considering how a student's learning experiences at school affected the ability to learn subsequently in a first year of apprenticeship, worryingly concluded that 'the relationship between experience of planning in science practical work at school and results of tests given at the beginning of training was a *negative* one: the more the experience of "planning" the lower the test score'.[18] As the test score related to planning skills of the ability to select, control and review methods appropriate to the successful completion of a given task, it appears that what 'planning' was being done in much school science practical was too convergent, too restricted, to be of value in devising and executing experimental investigations of a more realistic kind.

The third group of aims for doing practical work related to being a scientist, learning how to be a scientist by involving oneself in scientific activities. 'You can't be a musician without playing music' – obviously, but there is considerable doubt as to whether a lot of practical work being done by students in schools and colleges has got much to do with that done by 'real scientists in their laboratories'. There are fundamental difficulties here about whether practical work in schools can be of the same nature as real scientific investigation, and we will return to that in the next section. But the reality is that much, perhaps most, of the practical work done does little to develop, or reflect, the way a scientist works, although it may familiarise the students with apparatus and ways of working safely in a laboratory. 'Practical work in school science; it's certainly practical, but is it science?' This is a question that we must ask as teachers and curriculum developers as we plan yet another year's series of 'worksheet dominated, cookery book type of practicals'. The same stricture may well be applied to many of the laboratory activities provided for students in higher educational establishments.

The final group of aims centres on the Achilles' heel of school practical work – the persistent linking of practical to theory, with practical aims subservient to theoretical insights. Such practical work is designed to enable the student to discover and, subsequently, to understand more fully the theory being considered. Yet, somehow, students seem to 'discover' the wrong thing, that which seems so obvious to the teacher somehow eludes the student. Solomon has described how, at the end of an experiment which she had considered 'an almost perfect example of the guided discovery method at work', one of her students reproached her with the complaint that 'you haven't explained it'.[19] It was the teacher, rather than the experiment, who was expected to 'explain' nature! Very quickly the students learn the game of 'what are we meant to discover?', and the whole process develops into a highly structured exercise to ensure that the student *does* discover the right thing. Teachers have developed useful strategies to ensure that the experiment works; one such is articulated by Van Praagh, who starts by advising teachers 'give as precise instructions as possible regarding the conditions for carrying out the investigation'[20] and so students learn not to ask the wrong, i.e. their own, questions of practical work.

But it has been argued that the 'stage management' required to obtain the received answers is a crude sort of behaviourism in which children are deliberately treated by the teacher like rats in a Skinner box:

> Children will soon learn that the 'right' answer will turn up eventually and so they learn to avoid the intellectual stultification of putting forward their own 'wrong' ideas which are likely to be ignored, dismissed out of hand, or at best refuted by reference to evidence which was not available to them.[21]

There is an important distinction between 'discovering what' and 'discovering why', and it has been argued that only the former has much relevance to practical work.[22] The 'why' discoveries can only come from analysis, insights and hard abstract thinking on the selected appropriate data gleaned from practical observation. But even the 'what' discoveries can disappear rapidly in a mass of other observations, unless they can be fixed into some meaningful framework. Bruner quotes Miller in referring to 'the magic number of 7 ± 2', this being the maximum number of separate, unrelated items that most people can successfully retain in their short-term memory. It is possible for discovery learning to become 'rote discovery learning', when what is discovered does not engage with the student's previous cognitive framework. Probably the key to this problem lies in the 'noise factor' in practical work referred to by Johnstone and Wham.[23] The information overload bombarding a pupil doing a practical can be so vast that selection of the significant bits can become impossible. The 'organising information' game from the Science Teacher Education Project material, reproduced in Fig. 1.1 (page 12), indicates how difficult (if not impossible) it is for a 12 year old student to discover the order of the reactivity series of the elements from the mass of unrefined data derived from a series of experiments. So, if 'discovering' the theory does not work, surely practical work helps the student to understand the theory better. Certainly this has been one of the strongest arguments in favour of doing practical work; but now there is clear evidence to show that it may not be so, that many students, having done practical centred on some piece of theory, have thereby gained only a very slight, possibly faulty, understanding of the underlying theory. Tests of school students, of university science undergraduates and graduates, even of science teachers, show that they often have a seriously deficient understanding of basic theory. Driver has written, not entirely flippantly, that 'I do and I understand'

Fig. 1.1 The reactivity of some elements.

Source: **Science Teacher Education Project *Activities and Experiences*, McGraw-Hill (Maidenhead) 1974. This example is taken from Worksheet MT1.**

Here are a set of experimental results similar to those which a 12 year old pupil following the Nuffield Chemistry Course might have obtained in experiment A6.1. To help you see things from the pupil's point of view the names of elements have been changed. Suppose that you obtained these results from experiments in which you heated mixtures of metals and oxides on asbestos paper.

1 When exium is mixed with arium oxide and warmed, the mixture ignites and scatters all over the bench. You did this twice.

2 When zedium is mixed with arium oxide and warmed the whole mixture melts and forms a lump.

3 Elium warmed with arium oxide produces a puff of smoke, but it is not easy to see anything new formed.

4 Elium, arium and zedium seem to have no effect on exium oxide.

5 Arium and zedium seem to have no effect on elium oxide.

You didn't have time for any other experiments. Can you relate the elements one to another in order of their liking for oxygen, the most reactive first? What assumptions do you need to make?

should perhaps be replaced by 'I do and I am even more confused'![24] We have all met students doing practical work with ticker timers and trolleys who still do not understand the difference between velocity and acceleration, or working on the stability of model animals without any understanding of centre of gravity or equilibrium, or measuring combining weights of elements while hopelessly confused about interpreting the data at the particulate level. Of course much theory is difficult to comprehend, so we devise structured practicals to clarify it. But, as the A.P.U. conclude on the basis of their researches, 'despite the orientation of science courses to the teaching of content, the results from the tests of application of science concepts indicate that it is only a minority of 15 year olds who are able to draw on and use some of the *most basic* scientific concepts'.[25]

The more time one takes to listen and to hear what students really believe, the more convinced one becomes that doing practical work is an inefficient way of increasing children's theoretical understanding.

In conclusion then, we find cause for concern at the practical work being done in school science. Not all pupils find it interesting or motivating, the level of practical skills and techniques it develops is low, it is not realistic science, and it does not enable pupils to understand and master the underlying theory. Not a cheerful analysis! We are not, however, pessimistic; on the contrary, we see a vital and increasingly important role for practical work in science education and in the following chapters we will explore this.

But first we will look back a little at the way in which practical work has developed over the hundred years or so in which science has been taught in our schools, and see how the issues and tensions that we have touched on in this chapter have arisen and developed. Throughout it all we will see the hidden tension of contrasting views of education: the academic, cultural view of the curriculum as against the child-centred, useful view. In science teaching this has been expressed as a tension between stressing the importance of knowing pure science as part of our cultural heritage as against acquiring the habit of working like a problem-solving scientist in a relevant context, the appreciation of scientific knowledge as against the ability to do scientific tasks. Booth spoke of 'science the beautiful' and 'science the useful'.[26] All too often in our society pre-eminence has been given to the former. We believe that there is a need to shift the balance.

Notes and References

1 J. J. Thompson (editor) *Practical Work in Sixth Form Science*, Department of Educational Studies (Oxford) 1975.

2 J. W. Beatty and B. E. Woolnough 'Practical Work in 11–13 Science', *British Educational Research Journal*, Volume 8 (1982) pages 23–30.

3 H.M.I. *Aspects of Secondary Education in England*, H.M.S.O. (London) 1979 page 184.

4 G. W. F. Orpwood and J.-P. Souque *Science Education in Canadian Schools*, Science Council of Canada (Ontario) 1984 page 22.

5 R. J. Osborne and M. C. Wittrock 'Learning Science: a generative process', *Science Education*, Volume 4 (1983) pages 489–508.

6 H.M.I *Curriculum 11–16*, H.M.S.O. (London) 1977 page 28.

7 P. E. Richmond 'Who needs laboratories?', *Physics Education*, Volume 14 (1979) pages 349–350.

8 A. Jennings *Science in the Locality*, Cambridge University Press (Cambridge) 1985.

9 W. S. Toothacker 'A critical look at undergraduate laboratory instruction' *American Journal of Physics*, Volume 51 (1983) pages 516–520.

10 J. Kerr *Practical Work in School Science*, Leicester University Press (Leicester) 1963. Other attempts to classify the aims of practical work may be found in:

(a) G. Van Praagh 'Experiments in school science', *School Science Review*, Volume 64 (1983) pages 635–640.

(b) R. T. White, 'Relevance of practical work to comprehension of physics' *Physics Education*, Volume 14 (1979) pages 384–387.

(c) G. Gonzalez and J. Gilbert 'A level physics by the use of an independent learning approach: the role of the laboratory work' *British Educational Research Journal*, Volume 6 (1980) pages 63–83.
11 M. Polanyi *Knowing and Being*, Routledge and Kegan Paul (London) 1969. Also his *Personal Knowledge*, Routledge and Kegan Paul (London) 1958.
12 J. Ravetz *Scientific Knowledge and its Social Problems*, Oxford University Press (New York) 1971 page 15.
13 J. Kerr, see note 10 above, page 21.
14 D. P. Ausubel *The Psychology of Meaningful Verbal Learning*, Grune and Stratton (New York) 1963 page 144.
15 Assessment of Performance Unit (A.P.U.) *Science in Schools Age 13: Report No. 2; Age 15: Report No. 2*, Department of Education and Science (London) 1984.
16 W. S. Toothacker, see note 9 above.
17 H.M.I., see note 3 above, page 184.
18 Engineering Industry Training Board *School Learning and Training*, E.I.T.B. (Watford) 1977.
19 J. Solomon *Teaching Children in the Laboratory*, Croom Helm (London) 1980 page 9.
20 G. Van Praagh 'Experiments in school science' *School Science Review*, Volume 64 (1983) pages 635–640.
21 P. Stevens 'On the Nuffield philosophy of science' *Journal of Philosophy of Education*, Volume 12 (1978) pages 99–111.
22 J. J. Wellington 'What's supposed to happen, sir?' *School Science Review*, Volume 62 (1981) pages 167–173.
23 A. H. Johnstone and A. J. B. Wham 'The demands of practical work' *Education in Chemistry*, Volume 19 (1982) pages 71–73.
24 R. Driver *The Pupil as Scientist?*, Open University Press (Milton Keynes) 1983 page 9.
25 A.P.U. *Science in Schools Age 15: Report No. 2*, Department of Education and Science (London) 1984 page 190.
26 N. Booth 'What next?' *School Science Review*, Volume 61 (1979) pages 153–156.

2

100 Years of Practical Work

Nineteenth-Century Science Teaching

By the mid nineteenth century, science teaching was well established in some elementary schools, with laboratories and apparatus being provided from government grants. Enthusiasm for practical work may stem in part from the highly popular publicising work of famous scientists of the time, particularly Davy and Faraday at the Royal Institution.[1]

While elementary school science wilted under the constraints of the 1862 Revised Code, science in the public schools gradually gained a foothold in the curriculum. In general, the lecture demonstration was the preferred mode of instruction, being used for illustration and verification. The Schools Inquiry Commission of 1868 was sharply critical of the lack of science in many schools and provided an interesting early statement of what might be achieved through a practically orientated science course:

> True teaching of science consists not merely of imparting facts of science, but in habituating the pupil to observe for himself, to reason for himself on what he observes, and to check the conclusion at which he arrives by further observation and experiment.[2]

The Influence of Armstrong

Henry Edward Armstrong came to his great interest in school science from the background of a professor of chemistry. He was much influenced by developments in chemistry in Germany and by his own experimentation with unorthodox teaching methods with university students. His heuristic method is based on the notion of 'placing the pupils in the place of the original investigator'. He believed that students should themselves perform many of the experiments and that they should thereby discover for themselves the subject matter in the science course:

The use of eyes and hands – scientific method – cannot be taught by means of blackboard and chalk or even by experimental lectures and demonstrations alone; individual eyes and hands must be practised actually and persistently from the very earliest period in the school career.[3]

On account of his high standing as an academic chemist, Armstrong was able to use the British Association as an outlet for his persuasive arguments. At the British Association meeting in 1899 he described his ideal science course: there were six stages which are shown below. Only the fourth stage fully exemplifies the heuristic approach.

1 Lessons on common familiar objects.
(Involves observation, description, classification)
2 Exercises in measurement.
(Numerical measurements in a physical setting, e.g. volume, density)
3 The effect of heat on various elements and compounds.
(Observation and recording)
4 The problem stage.
(The most radical, with problems such as determining what happens when iron rusts, separating the active from the inactive constituents of air, determining what happens when sulphur burns, emphasising observation and hypothesising)
5 Quantitative determination of the composition of compounds.
6 Introducing theory, particularly the molecular and atomic theories.

With Armstrong's powerful advocacy, the heuristic approach rapidly gained supporters, under the watchful eye of *Nature* (1901):

A revolution which gathers strength every day. The system of science education by didactic methods still exists in places, but only because the machinery for carrying on the work on more rational principles has not been obtained.

The journal provided its own interpretation of the approach in 1904:

Two things are essential for Professor Armstrong's plan, first, that the pupils should perform experiments with their own hands, and second, that these experiments should not be the mere confirmation of something previously learned on authority, but the means of elucidating something previously unknown, or of elucidating something previously uncertain.[4]

During the early years of the twentieth century there was a rapid growth of schools taking up Armstrong's heuristic suggestions. Laboratories were appropriately equipped and teachers trained in the 'Armstrong method'. The guidelines from the Board of Education stressed the importance of individual work, along with complete and accurate recording of all observations, exact expression and correct inference. The emphasis throughout was on method of enquiry rather than subject matter.

By the end of the First World War, there had been a number of major criticisms of the heuristic method. The emphasis on laboratory exercises led to a bias towards measurement while neglecting other aspects of the pursuit of knowledge. Focussing on training in experimental method led to the neglect of the teaching of scientific principles and of science as a humanising influence. Teachers were giving their own emphases to Armstrong's course, perhaps finding the earlier stages (e.g. exercises in measurement) much easier to handle with large classes than the later, heuristic elements (e.g. the problem stage). Deficiencies in Armstrong's view of scientific method became clear, particularly in the inadequacy of his view that careful measurement of everyday phenomena constituted scientific method. The most powerful attack was mounted in the Thomson Report of 1918, which stated, after a thorough analysis of the heuristic method,

> We are driven to the conclusion that in many schools more time is spent in laboratory work than the results obtained can justify . . . insistence on the view that experiments by the class must always be preferred to demonstration experiments leads to great waste of time and provides an inferior substitute. The time gained by some diminution in the number of experiments done . . . could be well used in establishing in the pupils' minds a more real connection between their experiments and the general principles of science or the related facts of everyday life.[5]

In effect it was restricting what was taught in science lessons to those topics which could be studied practically by students in the laboratory. Discussion of the relative merits of individual experiments and demonstrations was not new but reached new heights at this time, fuelled partly by the critique of heuristic approaches and partly by the arguments about saving time produced by proponents of general science and science for all. As early as the 1920s, in response to situations in schools where there was insufficient apparatus for every group of pupils to perform the same experiment at the same time, many teachers had developed a worksheet system, called the 'card' system, where different groups were carrying out different experiments following instructions from a printed card – a circus approach.

Developments 1920–1961

Between the Thomson Report 1918 and important Science Masters' Association/Association of Women Science Teachers (S.M.A./A.W.S.T.) policy statements of 1957 and 1961, comment on practical work in official reports is generally restricted to debate about the relative merits of demonstrations and individual practical work. However, Part 2 of *The Teaching of General Science* listed, for the first time, pupil abilities which should be tested in examinations, including a section on practical skills.[6] This includes (a) the development of manual skills and dexterity, (b) the ability to do neat, accurate work, and (c) the ability to apply science to solve practical problems. This type of simple classification of skills was not developed again until the objectives movement of the 1960s. A further section of the report discussed the development of scientific modes of thought, specifically (a) the ability to explain the principles from facts and to support principles with facts, (b) to distinguish between fact and hypothesis, and (c) to plan experiments and draw conclusions. This provided a more rounded view of a scientific approach than that put forward by Armstrong.

The Spens Report of 1938 discussed experimental work in detail and criticised the waste of time in much of the then current laboratory practice. Against the background of the strengthening general science movement this appeared in the form of a strong commitment to more demonstration work, thus:

> a larger variety of experimental work, covering more ground and carried out more accurately and skilfully, will provide more data over a wider variety of topics than is possible where the only, or the main, experimental work is done by the pupils. By a greater use of good demonstration we believe that science teachers will more commonly stimulate wonder and imagination.[7]

During the late 1950s many interested parties expressed concern about the state of science education in schools. It was at this stage that the two strands were most clearly formulated which were to stay, largely separate, for the next 25 years. A watershed for future curriculum development had been reached. At this time secondary education in the United Kingdom was still sharply divided between a majority in secondary modern schools and a selected minority in grammar and independent schools. Two strands can be matched to the two systems. In 1953, and 1957, the S.M.A. published a report, in two parts, on *Secondary Modern Science Teaching.*[8] This pointed to a more child-centred, and less academic, science, where factors such as relevance and significance were important, and the students' practical work more investigational and open-ended. Subsequently, in 1960, Her Majesty's Inspectors of Schools published *Science in Secondary Schools*, which advocated science for all students throughout their secondary schooling and gave perceptive criticism and guidance concerning practical work. The report described critically laboratory work where the end is known, to the master at least if not the pupil, and the *modus operandi* has been carefully thought out and tested by the former so as to produce a certain recognised result. There is nothing, as a rule, to correspond to the clear formulation of a question by the pupil himself; this is provided for him and no value is placed on curiosity. Nor is there any necessity to construct a plan of investigation, to design and make *ad hoc* experimental devices or to modify them in the light of experience. Nor, again, if the answer comes out 'right', is there much inducement to consider the results, to estimate their validity or to discuss their further improvements. Finally, there is missing the ultimate satisfaction of having really found something out.[9]

The report advocated that pupils should carry out real personal investigations as part of their laboratory work. Not that these should replace entirely the demonstrations or practical exercises but that '. . . original investigations could be introduced into the practical work rather more frequently than at present'.

The reluctance to take on open-ended investigational work may be partly explained by considering the way in which practical work was to be encouraged for the minority of pupils in grammar and independent schools. In 1961 we find the Association for Science Education publishing a series of policy statements, under the heading of *Science and Education*, putting forward the view that 'all pupils should follow a balanced core of science subjects up to the end of the fifth-form year to ensure that those leaving school were properly educated or truly cultured and therefore able to participate fully in the life of their time'. These general principles were developed in three separate booklets for biology, chemistry and physics, a framework of content and practice being spelt out for each subject, leading to a course which was comprehensive, coherent and modern. The topics to be covered were laid down in some detail, with suggestions for experimental work. This practical work was to be linked closely to the theory, to illustrate or to derive it. Although some of the experiments were described as of a 'research or project nature', it was clear that such were perceived in a very convergent and theoretical way. For example, experiments to discover 'the laws of reflection of light' or 'whether the extension of a spring is proportional to the load', were classed as being of a research or project nature.

And so, around 1960, close to the beginning of the involvement of the Nuffield Foundation in the funding of curriculum development projects in science, the two styles of science teaching were clearly articulated. Which would be accepted? The knowledge-based, grammar school, 'beautiful' science of the Association for Science Education or the student-centred, modern school, useful science of Her Majesty's Inspectors? Not surprisingly, in view of the backgrounds of those who became involved in the early Nuffield developments, it was very largely the former. Although the first three Nuffield courses to be developed all stressed the need to develop in students a spirit of enquiry in order that they should have a critical approach to the subject, in practice the experimental work became convergent and content-led.

Practical Work in Nuffield Science

The first generation of Nuffield science projects was derived very closely from the frameworks provided by the S.M.A./A.W.S.T. policy statements. We will consider initially the Ordinary and Advanced projects in Biology, Chemistry and Physics, because they represent the first attempt in the United Kingdom at large-scale curriculum reform. Here follows an analysis, for each of the projects, of what has been written by the curriculum developers about practical work, and short commentaries on the responses to the use of materials in schools.

Nuffield Ordinary Biology

In this course it is stressed that experimentation and enquiry are to be emphasised ahead of 'mere factual assimilation'. One of the aims concerned with practical work is:

> to teach the art of planning scientific investigations, the formulation of questions, and the design of experiments (particularly the use of controls).

Four types of practical work are identified.

Class practical work is seen as either using information, techniques and concepts or working them out. It is strongly promoted as being an investigatory or problem-solving activity. The writers recognise that it is important for younger students to experience a sense of achievement in their practical work, and that this requirement may well be in conflict with the presentation of a genuine experimental situation which involves a degree of uncertainty. Thus rapidly do the authors draw back from genuinely investigational approaches!

Demonstrations allow the teacher control over the imparting of a piece of information, the introduction of a point for discussion, the description of a concept or technique. Without providing any, it is stated that the evidence available suggests that, for the above aim, the demonstration is as efficient as, if not better than, class work.

Group practical work is encouraged, where different groups are studying different aspects of the same theme, followed by a period of pooling and discussion of data.

Long-term investigations lasting two, three or more weeks are suggested for biological systems. However, the revised course acknowledges that there are serious problems in maintaining this work with large numbers of students and suggests the provision of data for analysis as an alternative in some cases.

Nuffield Advanced Biology

This course was developed end-on to the Ordinary level course and therefore some continuity of aims might be expected. The Advanced Biology course is firmly within the framework of science as enquiry, and stresses the importance of developing skills and techniques of experimentation.

Much emphasis is placed on the role of 'investigations' in the course which can either involve practical work or be based on second-hand data. The majority of the investigations 'have been carefully designed to provide reliable results' and carry instructions for experimental procedure along with questions that provide a guide for analysing the results. Should they be more properly called practical exercises? Extension work is given which is much more open-ended and which relies heavily on the student's initiative. Each student is required to complete a project which is formally assessed and which should occupy approximately 30 working hours. The topic to be chosen for the project should be simple, the apparatus required not complex.

Commentary on the Biology courses

While espousing the cause of enquiry approaches and investigations, those involved in both courses at the development stage recognise that there is a problem in providing open-ended investigations suitable for use in the ordinary classroom with whole classes, particularly with the pressure in the early years to obtain 'satisfactory' results from experimental work on which to build theoretical understanding. Kelly, writing of the Advanced level course, recognises the difficulties with practical investigations:

> The practical work in the Laboratory Guide has been carefully vetted and tested to ensure that it can be successful. Furthermore, the student is guided to some extent by lists of procedure and questions in the work. In these respects, then, it is contrived and does not provide a student with the experience of tackling a

genuine open-ended problem. A project, on the other hand, provides the opportunity by which students can individually gain this experience.[10]

Nuffield Ordinary Chemistry

The Chemistry writers clearly identify with the important goal of encouraging pupils to be scientific about a problem; of understanding Chemistry as being a way of conducting an enquiry. They argue that the basis for this enquiry must be firmly experimental. Experimental work is to be undertaken in such a way as to awaken the spirit of investigation and each experiment will normally involve three equally important stages: planning how to tackle the problem, carrying out the experimental work, and discussing what deductions can or cannot be made from the results. It is readily acknowledged that the experimental work presented to the pupils, and its outcome, will normally be well known to the teacher. The situation which is familiar to the teacher is taken as new to the pupil and the phrase 'guided investigation' is occasionally used in descriptions of the project by the organisers. The two major parts, or stages, of the course indicate a view of how science develops from concrete experience to theoretical insight thus:

Stage 1 (age 11–13) A time for 'doing experiments' in order to learn about a wide range of materials and some of their patterns of behaviour. Practical skills of measurement and observation are to be acquired and the beginnings of training in disciplined speculation.

Stage 2 (age 13–16) A time for attempting explanations in terms of particular models. These are developed only when they can be used. Practical work in this stage is very different and is slanted more towards testing theoretical models or ideas.

Nuffield Advanced Chemistry

The approach to practical work in the Advanced Chemistry course is a logical continuation of the approach of Stage 2 of the Ordinary level scheme. With the emphasis being put on a strong integration between theory and practical work, the strategy is to use theoretical principles to predict the outcome of reactions, then to test these predictions experimentally. A typical example is the use of $E^{\ominus}$ values to predict the outcome of redox reactions, prior to verification in the laboratory. This is a faithful interpretation of the view of practical work in the sixth form first developed in *Chemistry for Grammar Schools*.[11] The rgument presented is that sixth-form work should be an introduction the methods of intellectual exploration as much as an occasion for cquiring information and techniques. Practical work is to provide e basis of theoretical discussion and practical techniques are to be cquired, not as ends in themselves nor to satisfy examiners in practi-al examinations, but only as the necessity arises during some eoretical investigation. The requirements for the compulsory acher assessment of practical work provide guidance about practi-al work priorities. These are the ability to observe, the ability to terpret observations, the ability to plan experiments, and skill in anipulation and appropriate attitudes to practical work.

ommentary on the Chemistry courses

he role of practical work is clearly stated in the two programmes. he two link coherently into a progression from investigations lated to the study of materials to experiments more closely linked theoretical development in Chemistry. Nuffield Ordinary level hemistry has probably been more closely associated with the surgence of Armstrong's ideas than any of the other courses. It is, owever, significant that in none of the published materials nor in e connected writings of the team members at the time are words ke 'heurism' or 'stage-managed heurism' used. The closest phrase sed is 'guided investigation'. The writers are at pains to point out at they understand the limitations of student experimentation, articularly with respect to the time consumed (the familiar gibe at euristic approaches from the Thomson Report onwards). As one of e team members who came from a school with a direct link with the rmstrong tradition had already published a text with the title *hemistry by Discovery*, and was responsible for developing one of e Stage 1 alternatives, it was not surprising that the approach to xperimentation was given a heuristic tag. Van Praagh's own advo-acy of learning by discovery can be summarised as:

> . . a guided tour with some rewarding discoveries en route, rather than an undi-ected exploration.[12]

uffield Ordinary Physics

he original Ordinary level course has remarkably little to say by way f justification of laboratory work. The major argument is presented terms of students acquiring the feeling of doing science, of being a cientist. Students are asked to mirror the activities of the scientist in

devising their own experiments, meeting difficulties as well as suc
cesses, and trying things out with a watchful eye and a critical mind
Both student experimentation and demonstrations are encouraged
with the acknowledgement that student experimentation will take
long time and that perhaps the number of such experiments shoul
be restricted, with an associated strengthening of the role of demon
stration work. The writers appear to take almost literally the Brune
dictum:

> The schoolboy (*sic*) learning physics is a physicist, and it is easier for him t learn physics behaving like a physicist than doing something else.[13]

The revised course has only a little further discussion to offer regard
ing practical work, but there is one important commentary where th
writers defend themselves against the threat of their approach bein
labelled as 'heuristic' or called the 'discovery method'.[14] In the pro
cess, they succeed in parodying Armstrong's heurism but then do g
on to make a positive statement of their own intentions. They argu
that they do indeed want students to do their own experimenting, bu
with reasonable guidance and not under the illusion that they ar
doing new science. The urge for discovery and the delights of succes
are to be encouraged without deceiving students about history. Thi
stress on personal discovery, and the desire to develop in th
students a total reliance on the validity of their own results, wa
reckoned to be so important that no student texts were publishe
with the course. That, it was argued, would have undermined th
basis philosophy by allowing the students to turn to the text for th
'right answer'. If the absence of a student text enabled the student t
write freely, if insecurely, the form of the teachers' guides caused th
teachers to respond more prescriptively. Teachers found it difficul
to work from the often verbose and idiosyncratic teachers' guide
preferring to lean on the more specific guides to experiments. Con
sequently, some tended to follow the recipe without appreciating th
rationale, producing courses which were little more than a succes
sion of 'experiments to be done'.

Nuffield Advanced Physics

The experimental work is prescribed in some detail in the *Teacher Guide* and *Students' Guide* and justified in the *Teachers' Handbook*
Although it is admitted that there is probably not a 'best way' o
teaching the course, five particular types of practical work ar
suggested. *Demonstrations*, either as polished performances by th

teacher or as foci for discussion with student involvement, are included. For some topics a series of *related experiments* are provided, either suggested by previous ideas, or needed to open up a new topic. These can be done by different students who report back their findings to the class as a whole. A category of '*individual exploratory experiments*' is introduced where students are encouraged to decide for themselves what to do with some apparatus provided, not knowing what will happen. With many of the experiments being short and deliberately simplified, it is suggested that the students should also do one or two *long experiments*, giving the satisfaction of completing a task of some substance and of overcoming a number of difficulties on the way. To yield experience of the way of working of an investigational physicist, a fifth type of practical is included – completely open *investigations* to be undertaken on an individual basis. Here the students are expected to identify their own problems, invent their own (simple) experiments and deal themselves with the difficulties that arise. These investigations are seen as a fundamental part of the design of the course and aim to help students to become better at doing physics rather than to teach them more physics. Two weeks in each of the years are to be set aside for two individual investigations, the second of which is formally assessed. The investigations are included in order to provide the student with the opportunity to experience something more of the art of enquiry than is normally possible with the main, structured experiments in the rest of the course. By enquiring, the student learns more of *how* to enquire and may also begin to understand the sensitive relationship between theory and experimentation. Investigations are further justified in terms of the enjoyment and involvement evident in the way in which students tackle this aspect of their work, and the responsibility which is thus given to them.

Commentary on the Physics courses

Both courses brought together, or developed, a range of experimental work which was both elegant and ingenious and enabled students and teachers alike to share in practical work previously used only by the most innovative teachers. This was most welcome and refreshing to courses which had, for many, become arid and didactic. The Ordinary level course was perhaps too radical in its rather idealistic and ill-defined emphasis on a 'content-led discovery learning', an emphasis which needed perceptive teacher insight and commitment if it was to have any chance of success. The practical work was predominantly framed by the theoretical content that it was intended to illustrate or to discover and, although it was suggested that some experiments should be used to develop the investigational approach of 'being a scientist for a day', the examples given were generally convergent. It sought to provide 'physics for the enquiring mind', and all the physics that an educated student should know before leaving school, but this was, perhaps with hindsight, too academic for the needs of the majority of students.

The lack of an assessment of practical skills in the Ordinary level Physics course caused some blurring of the intended aims of practical work. Regular commitment to assessment can lead to clarification and reinforcement of practical work aims. In contrast, the Advanced Physics course provided more specific aims for practical work which were stressed in the examination process. The first part of the assessment was based on teacher assessment of work done by the students in their second investigation. The second part, designed to test particular experimental skills and techniques, was made up of a series of eight short exercises to be done in a single 90 minute examination. These new approaches emphasise again the importance of practical assessments in influencing the types of practical work being done. This is especially true of the Advanced Physics investigations where students have consistently shown themselves capable of achieving scientific investigational work of a very high standard.

Nuffield Combined Science

Nuffield Combined Science, the course which has provided the overwhelmingly popular basis for the development of science courses in the 11–13 year age range, breaks little new ground in its brief rationale for practical work – it effectively mirrors the approaches of the separate subject courses from which it was derived. As a consequence, different sections of the course carry the marks of the Nuffield Ordinary level materials. The familiar message is reiterated in the *Teachers' Guide*:

> In order to gain as much firsthand experience as possible of science as a method of enquiry, children must be attentively engaged in laboratory work. In this way children will develop an appreciation of how to formulate, test and modify hypotheses.

And again teachers were urged to allow students 'time to . . . design experiments and activities to test their suggestions . . .'.

Nuffield Secondary Science

The Nuffield Secondary Science course was developed as a pragmatic response to the needs of 'that part of the 14–16 year old population not provided for by the first generation Nuffield courses'. Because it was not confined by adherence to Ordinary level syllabuses, it was, to some extent, able to take new directions. The Newsom Report was sharply critical of much practical work in science, arguing that

> Too much of the tradition of science teaching is of the nature of confirming foregone conclusions. It is a kind of anti-science, damaging to the lively mind, maybe, but deadly to the not so clever.[15]

The project team was, therefore, insistent that the work presented must have *significance* for the students and must involve the *investigation* of *real problems*. In this there are echoes of Pamphlet No. 38 (Science in Secondary Schools) and the secondary modern tradition of science teaching. Idealism is, however, tempered with realism. The writers argue strongly that a greater element of investigation is crucial, while recognising that students who are not high fliers will need considerable help in posing the questions to be asked if their enquiries are to meet with adequate success. They accept that the use of open-ended experiments will have to be carefully regulated if confusion and depression from an apparent lack of progress and achievement are to be avoided. Alexander, in her evaluation of the course, found that teachers were providing a fairly high degree of support for the student through structured situations, while giving freedom within the framework. She found evidence of a move towards student participation, though not of a totally open-ended approach.[16]

The Implementation of Nuffield

Having produced impressive curriculum packages, the influential and often charismatic members of the Nuffield teams set out to sell their wares in an educational and financial climate that was receptive to such changes. But the steps from innovator's vision to classroom reality were not easy.

The Nuffield development teams had been very hesitant to produce syllabuses for their courses, preferring instead Sample Schemes and Teachers' Guides which became the reference points for those involved in developing new public examinations. In the case of practical work there was no such hesitation. Experiments were developed, tested and then made available to teachers in recipe format. Although always intended as samples, as suggestions for good practice, this new resource was taken by the teachers and built into their teaching without their having to question the purpose of a particular practical sequence. For the young and inexperienced science teacher to lean on, such guides to experiments were invaluable, until they became part of a new orthodoxy. This presentation of experiments in fine detail and in careful sequence was quickly reinforced by the commercial interests of publishers and apparatus manufacturers. The new ideas for experiments presented in the Nuffield course books quite rapidly became incorporated into commercial textbooks.

Apparatus manufacturers had been invited to respond to the new initiatives of the Nuffield teams, and provided prototype equipment for schools involved in trials of the new course materials. New apparatus was developed to meet the needs of particular new experiments, for example, smoke cells for Brownian motion, and gas syringes for a range of chemical reactions where collection of gases was involved. Some apparatus was offered in kit form so that a school could purchase a 'Nuffield materials kit' which, in addition to valuable items such as standard metal cubes of different dimensions, might also include plasticine and drinking straws, which are more generally available. In the second half of the 1960s, schools were thus able to fill their laboratories with apparatus which allowed whole classes to 'do experiments'. There was, therefore, a period of curriculum dissemination by apparatus rather than dissemination by, or of, a curricular philosophy. Undoubtedly, both publishers and equipment manufacturers were setting out to be responsive to the new ideas, and ending up by emphasising the inertia in the system. This was accentuated when specific, ear-marked local authority funding for Nuffield projects came to an end in the early 1970s and schools' buying power diminished dramatically.

Since the development of the Nuffield Science courses there has been effectively no significant new thinking about the type of practical work suitable for the 11–16 year age range. The new repertoire of Nuffield experiments has been widely accepted (though the particular investigational emphasis of the Nuffield Secondary Science Project largely ignored). Local Educational Authorities, publishers, and groups of teachers have each developed 'new courses', especially for the 11–13 age range, but these have largely accepted the examples and the philosophy of the Nuffield Ordinary level courses and applied them to contexts different from those which were originally intended. There is now clear evidence that much of the Nuffield Combined Science course is ill-suited to the intellectual capacities of classes of 11–13 year olds with wide ranges of ability.[17] The result is a curriculum consisting of a series of lessons 'doing practical' which keep students occupied for 80 minutes while evading the difficulty of many of the underlying ideas. Teachers were now faced with teaching science to a much wider range of abilities, often in mixed-ability groups, and were concerned to find ways of teaching this science to them. This led to development of 'individualised' work schemes designed to allow all pupils to attempt a course originally designed for the most able. Some teachers were too busy adapting courses, and coping with new problems to stand back and ask whether the type of practical was actually appropriate. On the other hand, much creative rethinking work was being done in the primary schools, through the investigational approach of the Schools Council Science 5–13 Project, but this has, to date, largely been ignored in the secondary schools. We will speak more of this later, for we believe that the secondary schools have much to learn from the best primary school approaches when it comes to allowing students to work as 'real scientists'.

The Effects of Nuffield

Having considered the development of the new science courses with their distinctive emphases on the encouragement of practical work, it is important to consider how far they have been accepted and to what extent they have affected classroom practice. We can get an indication of the uptake of some of the courses by considering the entries for public examinations. Even the most popular courses, Nuffield Physics and Chemistry at Ordinary and Advanced level, have attracted less than 20 % of the total entry for those subjects, while neither Biology course has reached the 10 % level. But it would be unfair to take examination entries as a valid reflection of the influence of the Nuffield courses on science teaching. The match between the aims of the course developers and classroom practice is often not good. Some people teach Nuffield courses using very didactic methods, some teach 'traditional courses' at least partly through 'guided discovery'. Eggleston, in his major study, found considerable variation in classroom practice: 'the picture which emerged was somewhat other than what might have been expected in the wake of Nuffield'.[18] In *Alternatives for Science Education*, the authors tried to analyse new curriculum projects against various criteria.[19] One of them was to judge whether first the intention and secondly the outcome of the practical work was to discover or to confirm. Of the eleven new courses listed, they judged that the intention of the practical work was to discover for eight of them, and to confirm for three. They further assessed that the outcome in the classroom practical was to discover for three of them and to confirm for eight! The writers suggest that, although 'true experimentation and discovery' is present in some pre-secondary schemes and genuinely experimental investigation in Nuffield Advanced level schemes with projects (Biology, Physics and Physical Science), 'other examples are not easy to find and many of the Nuffield courses have spawned contrivances designed to produce the "right" answer nearly every time'. The student is therefore lulled into the belief that something has been discovered as a result of turning a switch to start a device which cannot normally fail!

Of the impact of the Nuffield courses, Her Majesty's Inspectors of Schools had this to say:

> The Nuffield Science Projects have had a great influence on the introduction of practical work in science courses of all kinds. The pendulum has swung too far in some areas, however, with so much emphasis on class practical work that demonstration lessons are rare.[20]

Such comments appear to be very similar to those made by teachers and Her Majesty's Inspectors 20 years earlier, 'Plus ça change, plus c'est la même chose'!

And yet in many ways practical work in schools *has* changed over those 20 years. Much more class practical is done now, although resource constraints may already be affecting this as expensive apparatus bought in the boom days becomes superannuated and is not replaced. Genuinely open-ended, investigational projects, though rare, hold a central place in some of the best science in primary schools, in an increasing number of 11–13 science classes, in some excellent non-examination and Mode 3 CSE courses, [21] and in some of the Advanced level courses. Demonstrations seem to be a dying art. At a time when, rightly, teachers are re-examining their aims and practices, it is no longer sufficient to 'do practical, because doing practical is a good thing'. Each of our teaching strategies needs to be carefully justified so that our resources, including the most vital resources of the time and potential of students and teachers, are not wasted on time-consuming trivia.

Notes and References

1 A fascinating account of early science teaching, with many modern echoes, may be found in D. Layton *Science for the People*, George Allen and Unwin (London) 1973.
2 Quoted in P. Uzzell 'The changing aims of science teaching' *School Science Review*, Volume 60 (1978) page 7.
3 H. E. Armstrong *The Teaching of Scientific Method*, Macmillan (London) 1891 page 9.
4 Quoted in D. Thompson 'Science teaching in schools in the nineteenth century' *School Science Review*, Volume 37 (1956) page 305.
5 Report of the Committee on the position of Natural Science in the Educational System of Great Britain (Thomson Report) *Natural Science in Education*, H.M.S.O. (London) 1918 page 55.
6 Science Masters' Association *The Teaching of General Science, Part 11*, John Murray (London) 1938.
7 Report of the Consultative Committee to the Board of Education (Spens Report) *Secondary Education*, H.M.S.O. (London) 1938 page 250.
8 Science Masters' Association *Secondary Modern Science Teaching*, John Murray (London) Part 1 – 1953, Part 2 – 1957.
9 Ministry of Education Pamphlet No. 38, *Science in Secondary Schools*, H.M.S.O. (London) 1960 page 62.
10 P. J. Kelly 'Implications of Nuffield A-level biological science' *School Science Review*, Volume 52 (1970) page 273.
11 Science Masters' Association/Association of Women Science Teachers *Chemistry for Grammar Schools*, John Murray (London) 1961.
12 G. Van Praagh 'Experiments in school science' *School Science Review*, Volume 64 (1983) pages 635–640.
13 J. S. Bruner *The Process of Education*, Harvard University Press (Cambridge, Mass.) 1960 page 140.
14 Revised Nuffield Physics *General Introduction*, Longman (London) 1977 page 4.
15 Central Advisory Council for Education (Newsom Report) *Half Our Future*, H.M.S.O. (London) 1963 page 142.
16 D. J. Alexander *Nuffield Secondary Science: An Evaluation*, Macmillan (London) 1974.
17 M. Shayer and P. Adey *Towards a Science of Science Teaching*, Heinemann (London) 1981.
18 J. F. Eggleston, M. J. Galton and M. E. Jones *Processes and Products of Science Teaching*, Macmillan (London) 1976.
19 Association for Science Education *Alternatives for Science Education*, A.S.E. (Hatfield) 1979 pages 28–29.
20 Her Majesty's Inspectors of Schools *Aspects of Secondary Education*, H.M.S.O. (London) 1979 page 184.
21 D. Ingleby and L. Winspear 'Experimental studies – an alternative 16+ work unit' *School Science Review*, Volume 64 (1983) pages 773–778.

3

Towards a Rationale for Practical Work

The Student as Scientist

In moving towards our rationale and framework for practical work we want first to reassess our position regarding the nature of science and the way a scientist works and also the way that students know and learn about their world. Such a review will, of necessity, only be very brief but will show something of the stance we are taking towards science, knowledge and learning and will enable others to recognise the base from which our rationale grows. There will be times when the discussion moves freely between scientists and students, but this is deliberate. We, as do Kelly[1] and Driver [2], would see students as scientists in their natural way of working; each naturally motivated to explore their world and to seek to interpret it for themselves and then make sense of it. The fact that all students do not always show this motivation to act as enquiring scientists in school laboratories may indicate more about the artificial and inhibiting nature of school science lessons than about the students' lack of scientific potential. Our aim, therefore, to develop the scientist in the student should be seen as a general educational rather than a vocational one.

As we look at the nature of science we see two quite distinct strands. The knowledge, the important content and concepts of science and their interrelationships, and also the processes which a scientist uses in his working life. In teaching science we should be concerned both with introducing students to the important body of scientific knowledge, that they might understand and enjoy it, and also with familiarising students with the way a problem-solving scientist works, that they too might develop such habits and use them in their own lives. We believe that much teaching of science is ineffective because there is an inadequate understanding both of the nature of scientific knowledge and of the process of science.

Science teaching often appears to work on the assumption that scientific knowledge is objective, detached from the learner, and needs only good teaching to implant that knowledge into the learners' mind. The teacher is central, and directs the students, who fulfil a 'rat model' in a behaviourist exercise. We believe that this is unhelpful. We see the students essentially active in the learning process in which they are continually enquiring, testing, speculating and building up their own personal constructs of knowledge. It is only by personalising such knowledge that it becomes valid, meaningful and useful. And although there is in a sense public knowledge in science, students need actively to construct their own personal awareness and meaning. Osborne and Wittrock state the case clearly when speaking of the generative learning model of the pupil:

> The brain is not a passive consumer of information. Instead it actively constructs its own interpretations of information, and draws inferences from them . . . to learn with understanding a learner must actively construct meaning. The successful learning of scientists' ideas is as much a restructuring of the way learners think about the world as it is the accretion of new ideas to existing ways of thinking.[3]

Much science teaching is based on the assumption that a scientist works by a process of learning, and formally articulating, the basic concepts and principles of science and then consciously applying those principles to solve problems. Such teaching is devoted to the aim of giving students this high level of conceptual understanding so that it may be applied to new problems. We do indeed see science as essentially a problem-solving activity but do not believe that this is always achieved through an academic learning and understanding of a mass of fundamental concepts.

Tacit Knowledge

There is another form of learning about, of knowing and understanding, the world in which we live, to which in our academically dominated education system we give too little attention. It is the process by which we learned to walk, to swim, to ride a bike, which appears to by-pass the conscious processes completely. And it is learning which is never lost even over extended periods of time. This form of learning is very important in the development of scientists, who, through experience, 'get a feel for' or an 'awareness of' phenomena. When making a device, or solving a problem, they will 'know' what material to use and which lines of attack will work, not because they have developed a formal understanding of the properties of materials or the contents of the problem, but because they have developed a

feel for them; through experience. Polanyi speaks of explicit know
ledge and tacit knowledge, the explicit knowledge being articulated
and cognitively assimilated into consciously formed theories, while
the tacit knowledge is never consciously articulated but is acquired
directly through our senses and 'held' in readiness for more direc
application.[4] He argues that scientists rely very largely on tacit know
ledge in the way they work. Ravetz follows this line, and speaks of the
way that a scientist works 'as a craftsman'.[5] Schools tend to over
emphasise academic knowledge, assumed to be the foundation stone
for the pure scientist, and under-emphasise the instinctive, taci
knowledge acquired and applied by the engineer and in the 'persona
activity of a creative scientist'. In so doing, we fail to tap an enormous
pool of abilities among our young people. We need to emphasise the
value of both explicit knowledge and tacit knowledge in our teach
ing. One of the values of students gaining concrete experience
through practicals is the direct gain of obtaining first-hand know
ledge of the materials, of getting 'a feel for the phenomena'. This is a
valid end in itself and in our view does not need to be made subser
vient to the needs of obtaining an academic understanding. When we
return to 'problem-solving, investigational practical work', we wil
see that there are various strategies for succeeding here, not all o
them depending directly on an understanding of the underlying con
cepts. Though one strategy uses logical deduction from conceptua
understanding, others use this first-hand knowledge, this feel, this
experience, this intuition to enable problems to be attacked through
experience and an enlightened guess, trial and error, and imagina
tion. Perhaps more problems are solved through the engineer's 'feel'
than the scientist's 'understanding'.

Perhaps an example will illustrate this. A few years ago when one
of the authors was teaching a group of sixth-form physics students
an investigational project was developed with a local scientific estab
lishment of international repute.[6] A problem facing the scientists
centred on the mounting of superconducting magnets. The coils were
to be embedded in a resin to maintain their shape but, whenever they
were immersed in liquid nitrogen to start the cooling process, the
resin cracked. They needed to find a resin which would withstand
such changes in temperature without breaking up. Clearly the factors
which determined whether the resin broke or not were mechanica
strength, thermal conductivity and thermal responsivity, and we
decided that the students should devise experiments to investigate
these properties for a range of different resin types. This they duly
did and developed some splendid and highly satisfactory investiga

tions which yielded valuable data on those properties and how they varied with temperature. And so we were ready to solve the problem by applying our basic principles of physics to the problem in hand. But, on reporting back to the scientists, we found that they had already solved the problem. And how? They had made up a range of different resin types and by putting them in and out of liquid nitrogen found, by trial and error, which ones were strong enough!

Many of us on hearing a piece of music, whether we are familiar with it or not, find that we can quickly recognise the composer. We know the style; we have an ear for music. We do not necessarily understand the style, or know the patterns and structures used in writing the piece; we don't need to. We have acquired a feel for the music of this composer by listening to it over the years and becoming familiar with it. Our knowledge of it is tacit. Few would advocate that we cannot appreciate it until we have learned, and understood, the underlying music theory. While it is true that a few want to do so, it is not necessary for the majority. The parallel with doing science is clear. Much of our knowledge of and feel for scientific phenomena comes from steeping ourselves in first-hand experience of a range of scientific effects, properties and phenomena and thus we acquire the tacit knowledge which allows us to make scientific judgements. The parallels with literature are also clear. We learn to appreciate books and we write not by analysing the sentence structure clause by clause, but by immersing ourselves in books, by reading, and by practising the art of writing. 'Doing science' has much similarity with the realms of music, literature and art – we should recognise that the practice of being a scientist is both an art and a craft.

It is clear, therefore, that the science teacher's task is to arrange appropriate experiences for the student in a range of contexts, so that the student is able to build up personal constructs which will lead on to an increasingly meaningful learning.

Insights before Comprehension

A common belief about practical work is that students will gain an understanding through doing an experiment. Students will start by not understanding; they will collect data from their experiences, then they will make sense of all the information and thereby gain an understanding of what happened. Why does this not happen more often in our science lessons? Because the students need to have some perceptions, some model in their minds, before they can make sense of all they see. Observations are not neutral, objective events. The things they look for, the perceptions of what they see are determined by the cognitive framework that they bring to the observing. More generally, Solomon, in analysing why practical work with some students was a success while with others it was not, sums up her experience thus:

> Imaginative understanding was not a sequel to successful experiments. On the contrary, it was an essential prerequisite.[7]

This is a crucial pointer to why students gain little understanding from their experiments. When we, as teachers, look at circuit board experiments they all seem so obvious because we see, and understand, how they all fit into our cognitive framework, our model of how current behaves in circuits. Coming to a series of experiments without a model, the student will find it very difficult indeed to make sense and form a model of a whole collection of varied and apparently unrelated observations. It really would need a brilliant student, as it has brilliant scientists in the past, to discover from a series of observations what is 'going on' in a series of circuits and to acquire the appropriate concepts of current, voltage and circuitry. Moreover the situation is worse even than the student coming to the experiment with no model in mind. Very often a firm and apparently reasonable picture of the effect about to be studied in the laboratory will have already developed from everyday life. And it is through this, pre-scientific and limited cognitive framework that observations can be made and interpreted. Those preconceptions which students bring with them into the laboratory determine how they perceive experimental data. It is not yet clear how consistently students hold their cognitive models, though some believe that they are tenaciously held, and very difficult to change. Where a student brings such a conceptual model into a laboratory experiment it will be possible to fit the observations into that existing framework, whether it is appropriate or not. This problem is particularly acute in those areas of science in which a student has already gained experience. Before entering a science laboratory, students will have had a number of years of experience of learning to cope with such things as force, energy, gravity, electricity, heat and light. Many students have developed quite sophisticated models to make sense of the world around them, and they will not readily give these up under the influence of experiments in the laboratory which may, or may not, fit well into their existing framework. For instance, many students come to the topic of electricity with the firm belief that electricity gets *used up* as it goes round the circuit – therefore the current decreases. This notion will have been formed by learning that batteries go flat after some *use*, that household electricity bills have to be paid regularly because the electric cooker has *used up* electricity which was produced by the power station. It is all very reasonable. It may well be reinforced if the primary school has put across the idea that 'electricity is a form of energy'. It will take more than a few ambiguous experiments with a circuit board to alter this until the student has sufficient intellectual maturity to differentiate between electrical energy, electric charge, electric current and potential. As Driver points out, in holding to existing paradigms and in viewing the world through them, students are acting just like scientists.[8] Quoting Planck she says, 'new theories do not convert people, it's just that old men die'. We do not believe that we need to be quite so pessimistic with our students but we do need to be aware of the preconceptions that they bring into the laboratory and build on them. To use Ausubel's words, 'The most important factor is what the learner already knows. Ascertain that and teach accordingly'. So our task in making the practical work successful for our students is two-fold. First to ascertain and disentangle the relevant preconceptions the student is bringing to the laboratory, and secondly to modify them through discussion and possibly demonstration so that insights are taken into practical work which will enable a better sense to be made of what is seen.

Discovery and Rediscovery

We believe that problems have arisen because of confusion between discovery and rediscovery. The first generation of Nuffield courses, in varying ways, purported to be about developing a questioning and investigative approach, about 'being a scientist for a day'. But, where it was also suggested that the practical work should be done to give a greater understanding of the theory, there was a fundamental tension within the practical work which we believe was never resolved. Stevens has argued that 'there is a logical conflict between learning and discovering'.[9] Learning science means learning the accepted scientific wisdom, and necessitates, if it is to be done through discovery practical work, a process of 'rediscovery' of the accepted scientific insights. It becomes a closed, convergent exercise requiring tight control by the teacher. The process of scientific discovery is essentially more open and divergent, and yet if students are allowed to be involved in it they are unlikely to discover the deep insights which more mature scientists took long years to reach. This fundamental tension cannot be resolved as long as we insist that practical work shall be subsidiary to the theoretical scientific knowledge defined by a given syllabus and is needed only to support it.

The Distracting Clutter of Reality

We have argued that the imposition of theory on practical work has had a detrimental effect on the development of scientific investigational skills. Conversely, we now want to argue that the imposition of practical work on theory has had a detrimental effect on the development of cognitive understanding.

When our students are studying a scientific phenomenon, the experience of it will have an important part to play in its conceptualisation. The foundations of theoretical concepts lie in reality and students need to 'personalise' their knowledge for it to become meaningful. But it is also true that reality can have the effect of distracting them away from the 'simple' elegance of the underlying concept. This is so often the case with practical work designed to illustrate a basic concept or principle, when students get so enmeshed in the detail, in the measurements, in the incidentals of the experiment, that they lose sight of the underlying theory. The distracting clutter of reality presents a real hindrance to the search for patterns, whether students are trying to make sense of the world in which they live or of the experiments in the school laboratory. Johnstone and Wham describe the students doing practical as being in a 'state of unstable overload'. They suggest that learning is severely hindered when the working memory is overloaded with incoming data.[10] Evidence from the Assessment of Performance Unit tests shows that students can often deduce principles more confidently if they are not cluttered up by the scientific flotsam of experimental detail and often inaccurate data.[11] Many important scientific principles and concepts are elegantly simple; too often we hide them by experimental trivia. How often is the elegance of the conservation of momentum or of Newton's laws of motion lost in a welter of measurements from ticker timers or trolleys? Or an understanding of the elegant principles of pasteurisation lost in the lengthy procedures for properly replicating the experiments in the school laboratory? For our belief in these underlying principles has been derived not by practical experience but by theoretical argument. Of course the original workers had to carry out experimental work but lengthy replication studies may actually be harmful. Let us therefore not be afraid of teaching basic theory clearly through demonstration and discussion, utilising the previous experience that the student has acquired of that concept. It has been said that the law of the uniformity of nature could never have been discovered in a science laboratory – that view might not have been entirely without substance. Eddington expressed a similar sentiment when urging us 'not to put too much confidence in experimental results until they have been confirmed by theory'!

Science deals with theoretical concepts and their interrelationships. They are abstract and have to be considered and manipulated in the abstract. It is essential that these concepts are separated from their concrete reality if the maturing scientific mind is to gain mastery of them. We mislead and restrict the thinking of students when we give the appearance of relating everything to a laboratory experience. (How many physicists, for instance, believe that momentum has some physical, as distinct from only a mathematical, reality?) Shayer and Adey have warned us of the dangers of trying to teach abstract concepts to students who are thinking concretely.[12] But the obverse of this is true, that we waste time and restrict thinking by teaching abstract concepts through concrete practical experiences to cognitively mature students who have already mastered the rudiments and basic vocabulary. Perhaps the logical solution to the problems of teaching abstract concepts to young students who are still capable of thinking only in concrete terms is not to attempt it! The effect is to reinforce or introduce misunderstandings in the students' minds which will take much unlearning later. If the concepts can be introduced at a later stage, when the student can cope with them in an appropriately formal way, they will be better appreciated and will not need to compete with misunderstood descriptions of concepts which were introduced formally too early. So let us consciously remove the formal, abstract ideas of science from their practical base in order to learn how to handle them maturely.

Cutting the Gordian Knot

The first step we need to take is to deliberately and consciously separate practical work from the constraint of teaching scientific theory. We must stop using practical as a subservient strategy for teaching scientific concepts and knowledge. There are self-sufficient reasons for doing practical work in science, and neither these, nor the aims concerning the teaching and understanding of scientific knowledge, are well served by the continual linking of practical work to the content syllabus of science. We believe that the common notion that practical work is used primarily to discover, or to aid understanding of, the theoretical content of accepted science must be changed. The two aspects of science, the content and the process, have become inextricably mixed in our science teaching. We believe that a beginning should be made by releasing practical from the shackles of theory, and attending to process as a separate objective, important in its own right. We will make no progress until we have cut this Gordian Knot.

There are of course important links that properly can, and should, be made between practical work and theory but, in the first instance, it may assist in clarifying our thinking if we separate the two.

Notes and References

1 G. A. Kelly *The Psychology of Personal Constructs*, Norton (New York) 1955.

2 R. Driver *'The pupil as scientist?'* Open University Press (Milton Keynes) 1983.

3 R. J. Osborne and M. C. Wittrock 'Learning Science: a generative process' *Science Education*, Volume 4 (1983) pages 489–508.

4 M. Polanyi *Knowing and Being*, Routledge and Kegan Paul (London) 1969.

5 J. Ravetz *Scientific Knowledge and its Social Problems*, Oxford University Press (New York) 1971 page 75.

6 B. E. Woolnough 'School–research laboratory liaison' *Physics Education*, Volume 7 (1972) pages 401–406.

7 J. Solomon *Teaching Children in the Laboratory*, Croom Helm (London) 1980 page 9.

8 R. Driver, see note 2 above, pages 9 and 10.

9 P. Stevens 'On the Nuffield philosophy of science' *Journal of Philosophy of Education*, volume 12 (1978) pages 99–111.

10 A. H. Johnstone and A. J. B. Wham 'The demands of practical work' *Education in Chemistry*, Volume 19 (1982) pages 71–73.

11 Assessment of Performance Unit *Science at Age 13*, Department of Education and Science 1984 page 28.

12 M. Shayer and P. Adey *Towards a Science of Science Teaching*, Heinemann Educational Books (London) 1981.

4

Aims and Approaches for Practical Work

Fundamental Aims

We wish to claim that there are three fundamental aims, central to the nature of scientific activity, which fully justify the use of practical work. Though these aims are specifically and distinctly related to an education *in* science, they are not exclusively so. Each has a more universal utility and so our argument would be for their relevance and inclusion in the general education of all rather than for the vocational training of a few. The three aims relate to:

(a) developing practical scientific skills and techniques;
(b) being a problem-solving scientist;
(c) getting a 'feel for phenomena'.

We will consider each of these and discuss ways in which these aims can be met.

Developing Practical Scientific Skills and Techniques

There is a range of practical skills and techniques that scientists have to acquire before becoming masters of their craft. The aim of developing such skills is fundamental in scientific education as 'one cannot be a craftsman unless one can manipulate one's tools'.[1] Central to the skills that need to be developed are those of observation, measurement, estimation and manipulation. It is important to acquire the ability to observe carefully, honestly and perceptively, to recognise similarities and differences, to appreciate what is significant and to be able to measure a variety of properties. The use of scientific instruments follows, enabling observations and measurements to be made of properties outside the unaided range of sensitivity of the human senses. And building on the observation and measurement skills are those of estimating values for physical quantities and making sensible approximations. Manipulative skills need to be developed to handle apparatus and equipment safely and appropriately. Other experimental work will be needed to develop appropriate experimental techniques to familiarise the students with, for example, the making of aspirin to show some techniques of organic chemists, the growing of plants under controlled conditions to indicate the techniques of horticultural biologists, the use of the cathode ray oscilloscope to introduce the multitudinous uses of this apparatus by the practising physicist or engineer. Furthermore, the techniques required to plan, to execute and to interpret the results of experiments must be developed. The ability to manipulate, and make sense of, the data from a practical experiment, and to appreciate the extent of its reliability, can only be acquired through practice. All of these skills of a practising scientist are important and need to be consciously developed through individual practical work.

While postulating the aim of developing specific scientific skills we believe that this should be seen as subservient to the second aim for practical work, that of developing the whole process of working like a scientist. This involves much more than merely having acquired mastery of a range of skills and techniques. There is another approach to the development of scientific skills which seems to us unhelpful and contrary to the spirit of scientific enquiry; we would not wish what we have just said to support it. That approach might be called 'atomistic' and, in its desire to teach specific skills and techniques and assess the degree of their attainment by pupils using predetermined criteria, can reduce science to a series of tiny, trivial, disconnected tasks. Moving rapidly from the broad processes of planning, performing, interpreting and communicating, to more specific skills of observing, measuring, manipulating, etc., some would wish to move into further detailed skill descriptions, such as observing similarities and differences between two insects, measuring temperatures to the nearest degree, manipulating a bunsen burner. Even these sub-skills might be considered too broad and might be divided into even more finely prescribed minutiae. Now such sophistication may be useful for diagnostic purposes, but when it becomes the driving force for the curriculum and for the teacher, it reduces science teaching to the mastery of many trivial tasks. Scientific activity as a whole is greater than the sum of the parts. So, for science teaching, we must ensure that it maintains its freedom, and its wholeness, and that what started as a useful support for developing practical skills and processes does not become a restricting framework. In the preface to the Revised Nuffield Ordinary Physics, Rogers wrote,

It is fashionable in Europe now to carry out a meticulous analysis of separate objectives and outcomes of teaching and learning, so that they can be assessed in ests. This 'taxonomy of educational values' grew in the work of Bloom and thers in the United States twenty years ago. As it developed it was a valuable evolt against careless, vague planning and testing. But it concentrates attention n aspects that are clearly measurable and it misses some of the most important actors in our hopes for lasting benefits from Nuffield science – the enjoyment, mbition, pride, that we look for in 'wonder and delight and intellectual satisfac-ion'.[2]

Being a Problem-Solving Scientist

Iaving seen the need to develop scientific skills (and also the dangers of this becoming an end in itself), we move to the central, holistic aim of practical work, that of developing in the student the habit of work-ng as a scientist. Primarily a scientist works as a problem solver; so should our students. They need to be given, or should suggest for hemselves, a problem in a scientific context, be encouraged to analyse the problem and decide what are the relevant parameters. They need to formulate ideas which can be tested and developed ater. They should learn to devise a range of possible lines of investi-gation and select the optimum track. They should then execute that nvestigation and evaluate their findings, modifying their procedures as necessary. This is the problem-solving approach of a scientist, closely allied also to the process of technological investigation. In a lifferent context[3] we have used the mnemonic PRIME to help pupils emember the stages:

P for Problems to be tackled,
R for Research into the appropriate factors,
 for Ideas about ways of attacking the problem,
M for Making the device or experiment,
E for Evaluating the outcome.

This problem-solving approach is open-ended, divergent, and is unlikely to be appropriate for the 'discovery' of a certain, predeter-mined fact or theory. In such work there will be no right answers, hough some solutions will be better or worse than others. It is far emoved from the type of 'problem-solving or research problems' suggested in the early 1960s, where the 'problem' was an essentially convergent, theory-led one, aimed at 'discovering', say, the laws of eflection of light, or factors affecting the period of a pendulum. The aim is to familiarise students with the approach used by 'real scien-ists', so that the activities in the science practical class are more like real science'. We want students to learn to work as a scientist, or an engineer, does – we are not interested in trying to get every student to replicate, in an hour or so, the discoveries which have taken past scientists years to develop. Freedom needs to be allowed for lateral open-ended thinking, and investigational, problem-solving project work can allow for this. As Medawar says, 'creativity cannot be learned perhaps, but it can certainly be encouraged and abetted'.[4]

It has been argued that scientific and technological problem-solving is achieved by applying a formal understanding of the underlying scientific theories. We doubt the validity of this view and believe that techniques of problem-solving can be developed by students (and scientists) using tacit knowledge directly, to a large extent, without conscious recourse to theory. Indeed, for some, working within a tight theoretical framework can prove to be limiting and restrictive.

One of the most constructive and clarifying influences on science teaching in Britain has been the work of the Assessment of Performance Unit (A.P.U.). In deciding on what aspects of the students' scientific performance they were to assess they defined the essence of scientific activity with six categories, culminating in the performance of investigations.[5] They speak of 'science *as* problem solving' and interpret this as a problem-solving chain (Fig. 4.1). This is the culmination of the craft of a scientist, and the ability to work in this way has vast educational benefits to all students – not just potential scientists.

There is, as a sub-set of the problem-solving approach of the scientist, a certain specific type of hypothesis testing of theories (i.e. the copper problem)[6]. We believe that this has an important place in science teaching, though not the artificially dominant place that it has assumed in some courses. Nor should it be a process approached exclusively through practical work; in many instances hypothesis testing can be developed more appropriately through data analysis, where the data are provided apart from the distracting clutter and inevitable inaccuracy of student practical work. The techniques of hypothesis testing are particularly suited to the computer, where many types of hypotheses and models can be tested quickly and confidently.[7]

Getting 'a Feel' for Phenomena

The third aim of practical work to which we attach importance is that of enabling students to acquire 'a feel' for the phenomena they are studying. Science is about getting acquainted with the physical world we live in, and making sense of it, so clearly our students should get 'a feel' for the phenomena of which it consists. That feeling may come through any of the senses or it may come through

Fig. 4.1 The problem-solving chain.

Source: **Assessment of Performance Unit *Science in Schools: Age 15 No. 2*, Department of Education and Science (London) 1984, page 82.**

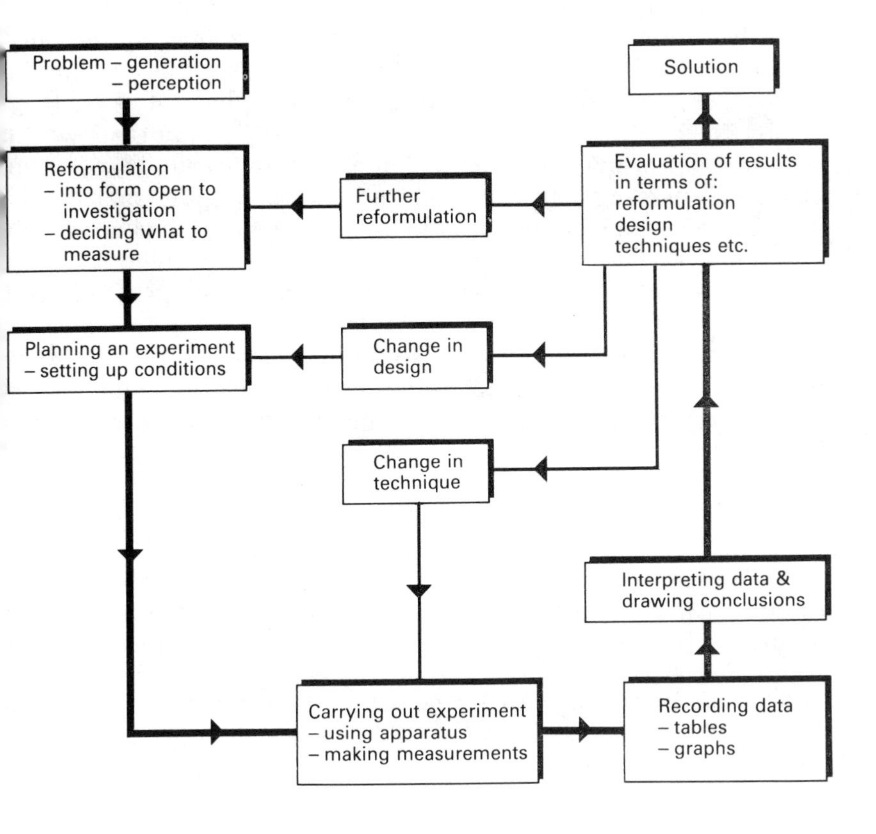

instruments. But however it comes, students need to appreciate a feeling for the world they are studying. They need to feel the force of an elastic band when they stretch it, to see and smell a gas as it is produced, and wherever appropriate to handle living organisms. We do not do a service if we produce scientists in the schools who have mastered the theory, can pass examinations and play clever scientific tricks, but don't appreciate the colours of soap bubbles, the growing alum crystal, or the relative sizes of an amoeba and a bacterium. In acquiring a knowledge and appreciation of the physical world they live in, they can, as the Victorians used to say, *enjoy* their world more. But knowledge of their world, for its own sake, is only one reason for ensuring that students acquire a feel for phenomena. It is also important for students to build up a reservoir of tacit knowledge which may be tapped directly when undertaking problem-solving. The reason that a scientist 'senses' the appropriate materials to use and seems instinctively to know what is likely to be the most fruitful line to follow is because a fund of tacit knowledge has been accumulated through past experiences which can be used even though it has never been formally defined. The art, and the craft, of a scientist can only be developed through practical, hands-on experience. Experience is also a necessary precursor to understanding the theoretical concepts behind the phenomenon, and can give reality to models and theories after they have been introduced.

Obtaining the knowledge through first-hand experience builds up a more meaningful grasp than can be acquired through theoretical argument alone. Speaking of practical work, a student recently said to one of us, 'when I do it for myself I really believe it!' – perhaps we should replace 'I do and I understand' by 'I do and I believe'. Students acquire tacit knowledge at first-hand and can then use it directly. Getting a concrete 'feel' for the phenomena being studied will be a necessary prerequisite for acquiring a theoretical understanding of the underlying concepts later. As we have argued earlier, such understanding needs tighter discussion and argument than can be provided in normal small-group practical work. It is important to make this distinction between getting 'a feel' for phenomena and getting an understanding of the theory. Practical work is important, indeed essential for the former, but we believe it is largely inappropriate to the latter. But, and it is here that the practical work re-unites with the theoretical knowledge, it will be necessary to appreciate, to have acquired 'a feel' for, the phenomena before the theoretical concepts can become meaningful. So the types of phenomena to which students are exposed will be determined, in part at least, by the theoretical concepts which are important for the student to master. Most students, however, will be able to use the 'feel' for the phenomena they have personalised in many types of problem solving, even without having formally understood the underlying theoretical concepts.

Matching Practicals to Aims

In discussing our three main aims for practical work in science, we have seen that they are quite distinct. It is not surprising, therefore, that we would suggest that there need to be different types of practical to fulfil the different aims. These would correspond appropriately: *exercises* to develop practical skills and techniques, *investigations* to provide opportunities to act like a problem-solving scientist, and *experiences* to obtain a feel for phenomena.

EXERCISES	Developing practical skills and techniques
INVESTIGATIONS	Being a problem-solving scientist
EXPERIENCES	Getting a feel for phenomena

Exercises

Exercises are designed to develop the scientific practical skills of observation and measurement, of manipulation of equipment and the various techniques that scientists use in their work. It is no longer possible to hope that students will pick up these skills *en passant* as they follow their content-dominated practical work. There is now a great deal of evidence to show that it just does not happen, because students are too concerned about 'getting the right answer' to worry about developing their skills.[8] We suspect this is largely due, again, to the distracting dominance of the theoretical content, where the teacher is unduly concerned with the students observing, say, the right thing. The students soon spot the game of learning to see what the teacher wants them to see. As they learn to answer the teacher's 'closed questions' by guessing 'what answer the teacher wants' (and not bothering to think about the question), so they learn to respond to the teacher's 'closed observational task' by 'seeing what the teacher wants them to see' (and not bothering actually to observe carefully what they are looking at). We would argue, along with the A.P.U., that opportunities should be arranged to give students practice at these particular skills in situations where the content itself is not the prime consideration. There is a need for practical exercises where it is clear, to the student as well as the teacher, that the *process* is more important than the *content* and that the purpose of a particular practical exercise is, say, to develop observational skills. Good, careful observation is a hard discipline and must be encouraged specifically. The CHEM Study exercise involving observation of a burning candle is something we can all learn from.[9] There is much that happens of note when we heat chemicals, even when we heat a beaker of water, and such simple tasks make excellent exercises – when we stress that the prime aim is to develop observational skills. The natural environment provides a splendid 'laboratory' for developing such skills. Many have gained enjoyment and a sense of wonder, as well as a deeper understanding of the order and patterns of the world, simply by learning to observe carefully the world around them. The natural sense of curiosity of young students is often lost as we allow the content to dominate and direct the students to observe only 'the right thing'. We must consciously revive the skills of observing differences and similarities, and of observing patterns, and the art of being able to appreciate and enjoy what is being observed.

Often the exercises developing such skills will lead straight into subsequent investigations and experiences, and may even be mixed in with parts of the whole experiment. We would certainly not want exercises to revert to those sterile practices detached from any reality and purpose. It may nevertheless be useful to separate, at least notionally, the different aspects of practical work to ensure that due emphasis is put on each function. Too easily the purpose of the investigation or the experience can be lost through inadequate skills, and the skill is not developed because of undue emphasis being given to 'the right result'.

The skills of measurement, with which we would put estimation and order of magnitude measuring are also fundamental. Simple, direct measurement of physical properties of length, mass, volume, and temperature can be linked with estimation 'games', and applied to material of direct interest to the students themselves – parts of their bodies and their belongings. Students have an intrinsic interest in these. Simple exercises need to be developed to bring initial familiarity with measuring instruments which can then be used for student investigation. Often students lose their way in an investigation because they become bogged down in mastering the details of the apparatus. Measurement of length can soon develop into investigations into human dimensions of span, stride and height. Use of the and lens and the microscope can soon lead, after careful, initial, irect observation, to pupil investigation of the life in a pond or part f the school environment. Such self-justifying uses of measurement re far more appropriate than artificial exercises aimed at, say, discovering the principle of moments by measuring the distance of brass iscs along a balanced lever – an aim so artificial, optimistic and nappropriate that it is surprising to find it still in so many science ourses. Measurement of weight, or mass (for all but the scientifically ophisticated student the distinction is a confusing irrelevance!) can oon be extended into an investigation such as that to find the energy vailable from the combustion of a peanut. Exercises with the splendid nicrobalance, which of all instruments helps to develop careful nanual skills, can quickly lead into simple investigations.

The measurement of derived units – density, specific heat capacity, acceleration – may well be useful after such concepts have been omprehended (through experiences). Then with the specific aims vertly stated, they may be used as exercises in measuring accurately r checking estimations or as an integral part of an investigation. Measurements of weight or volume to derive densities may be done fter students have obtained a feel for density, either as an exercise in ccurate measurement or as a check for exercises in estimation. The se of a ticker timer to measure motion and acceleration can be nastered through specific exercises, and then can be used in an nvestigation of, say, a student on a bike, a sprinter, or a weight lifter. Jsing ticker timers and trolleys to discover Newton's second law of notion is indeed using a sledge hammer to crack a nut, and a very onfusing sledge hammer at that. Most students, after they have ained a feel for force and acceleration through experience, find it lindingly obvious that the acceleration of a body is proportional to he force applied (especially when shown a simple demonstration) – t only becomes difficult when we wrap it up in complicated experimentation and calculation.

Before leaving the discussion of the need for exercises for students o develop the techniques of using scientific apparatus, we would ike to put in a plea that such apparatus should be appropriate, useful nd relatively modern. There can be little value nowadays in learning how to use Searle's bar apparatus or his apparatus devised to neasure Young's modulus for a steel wire. These pieces of apparatus, plendid and ingenious as they were when they were devised in the atter part of the nineteenth century, do nothing to give students 'a eel' for thermal conductivity or elasticity and can hardly be justified s providing students with useful training in contemporary experi-

Examples of Exercises

Observing and describing
- a candle burning
- water boiling
- the effect of heat on different chemicals

Using eye, hand lens, microscope
- to study the structure of paper
- to study the structure of fibres
- to study the structure of onion cells

Observing, drawing and classifying flora and fauna

Using keys

Measurement of commonplace and personal dimensions
- length, area, volume, weight

Estimation of dimensions and other measures such as
- room size and volume
- number of leaves and weight of a tree
- weight of air in a room
- number of molecules in one breath

Handling and transferring chemicals safely

Heating liquids and solids in test-tubes and beakers. Recording temperatures

Manipulating glassware safely

Working safely with living things
- vertebrates
- invertebrates
- micro-organisms

Using scientific equipment appropriately, such as
- chemical balance, measuring cylinders, burettes, poopers, microscope[s] vernier scales, timers, ammeters, voltmeters, cathode ray oscilloscopes

Correctly performing standard chemical tests
- for gases
- for starch, carbohydrates, reducing sugars
- for pH
- for important ions in solution

Developing, printing and enlarging photographic paper

Setting up correctly
- electrical circuits
- control experiments in plant growth
- apparatus for organic synthesis

mental skills. Sadly there is still too much experimentation done i[n] schools and higher education with apparatus which hides the under[-] lying science and which is itself no longer relevant. We worry tha[t] the current desire to get 'modern' apparatus into schools to prepar[e] students for the world of work will meet the same fate. With technol[-] ogy changing so rapidly, the schools will again be left behind wit[h] outdated equipment. We suspect that the development of skills wit[h] such specific, vocationally directed equipment is best left until the student leaves school and is 'on the job'.

Investigations

Just as problem solving is at the heart of a scientist's approach to working, so investigations must be at the heart of the practical work done by students. Investigations are designed to give students practice, and consequently the opportunity to develop competence, in working like a real problem-solving scientist. Investigations may come in different forms: they may be as short as half an hour or as long as half a term, they may be done by individuals or by groups, in class or at home, they may be related to the scientific content being studied in the syllabus, either leading into it or derived from it, or they may be independent of any specific content theme. But all will have the common characteristics as described in the problem-solving chain earlier. All will start with a problem, or a question, which is real to the student. How can I make cress seed grow most prolifically? Which adhesive, under what conditions, makes the strongest joint? What factors affect the rate of erosion of limestone? The question is most likely to arise from the science topic being studied, or it may arise from the students, or the classes, own interests. Having established the question, the student needs to analyse the factors relevant to the question and assemble any appropriate information. Various ideas and lines of attack need to be created and considered, the best one selected and the investigation planned. Many such experiments will use only simple, standard equipment, though some will require the student to make up a piece of apparatus to meet his own needs. As the investigation is executed, observations made and measurements taken, answers to the problem will be suggested. This will lead to evaluation of the experiment and a modification of the technique. In practice most students find, as do 'real' scientists, that this process is not linear but interactive and developmental. Planning, for instance, is not done effectively before the experiment starts but is a continuous process of the student working as a scientist with the experiment. There is much advantage in these investigations being worked collaboratively, either in pairs or with a group working around a common theme, as the interaction and feedback between the students can be most constructive.

In some areas investigational work is already well established. At Advanced level, the Nuffield courses in Biology, Physics and Physical Science all require completion of a project, and other courses allow ‘projects’ as an optional extra or alternative. Much field work in Biology is presented in an investigational mode. At the other end of the age range there are splendid examples of investigational work coming from children in primary schools. The Association for Science Education publication of children’s work in primary science illustrates something of the flavour and quality of such work.[10]

In the early and middle years of secondary education, however, investigational project work is still quite rare. Some Mode 3 science courses incorporate a component of teacher-assessed project work which may be of an investigational practical nature and which we hope will be retained in the new style examinations at 16+. It is sad that such exemplary work of a real scientific nature is often reserved for the less able students while the ‘more able’ are kept to a diet of sterile, convergent, academic knowledge!

More recently there have been examples of investigational work being introduced into the science courses of 11–13 year old students as teachers have rethought what is, and is not, appropriate at this level and decided they have more to learn from the student-based investigations of the primary school than from the syllabus-based curriculum of the potential entrant to high education. We have been working on a project with science teachers in the Oxford area and have found a ready response from teachers and students to the introduction of investigational work for this age group.[11] Investigations may arise from the normal topic being studied, are usually very straightforward and cheap, and may involve all the class looking at similar problems: investigating the factors that allow a simple parachute to fall in a controlled manner, how to prevent heat losses from a cup of coffee, how to make an exit sign which is visible at a long distance, finding the conditions for growing mung beans most prolifically, or investigating the composition of the soil around the school. Some investigations may be even more task orientated and are familiar through the ‘Great Egg Race’ type of competition: make a vehicle, driven by a standard elastic band, which carries an egg 20 m in the shortest time; construct a bridge out of straws/paper/balsa wood, which spans a 30 cm gap and supports the largest weight; make a paper bag as cheaply as possible which will hold 20 marbles but will break with 30. Other teachers have found that even less guidance provides students with greater scope for creativity and personal motivation, allowing a two to three week block of science lessons for students to make and develop a device to solve any problem of their own choice. After a short time students understand what is a realistic scale of problem to attack.

The outcome of such investigational practical work is varied and very real. The students obtain a sense of satisfaction and achievement from producing something, by themselves, which solves 'their' problem. Their commitment to the problem is very strong and leads to a determined involvement which is uncommon when doing the 'teachers' practical'. The quality of work done by students when their investigations are based on their own open-ended problems is significantly greater, indeed for many teachers surprisingly higher, than that more commonly produced by students in response to the teacher's more convergent practical task. Such investigations are able to encourage and develop in the students different and often unexpected talents of originality, creativity, independence, and help – incidentally – to develop those affective aspects of self-fulfilment, self-confidence, perserverence and commitment which are so important in the wider aspects of a general education.

The learning and knowledge that is gained in this way is often very deep, as it becomes 'personalised' around a personal problem. The cognitive framework is directly affected and modified as the feedback between the student's existing knowledge, the questions which arise, and the continually modified investigations, interact and lead to meaningful learning. The involvement of the student at all stages of the investigation leads to a personal acquisition of the new knowledge which goes very deep. No student who has investigated the shapes of milk splashes, and devised a method for photographing them at different stages through their formations, will ever pour milk into a cup of tea again without a renewed fascination for the phenomenon![12]

Perhaps, at this stage, it would be appropriate to quote at some length from a paper by Reay on 'Scientific investigation in our science teaching'.[13] She describes, in a way we could not better, her experiences in teaching science and the effect that introducing one type of investigational project had on her students.

A few years ago, I had a fifth form whom I had consigned two years previously to C.S.E. since they were not considered G.C.E. material. I tried very hard with them, giving them a great deal of practical work, worksheets, audiovisual stimulus and everything else that is supposed to motivate children to learn physics. No joy! They drifted lethargically to the lab and some of them went through the motions but their work was mediocre, their practical books were a mess and there was too much play for my liking.

One day. after a simple activity with circuit boards the class drifted off, not even stealing any components (which in itself was evidence of lack of motivation). However, Dan stayed behind and asked 'What did we learn in that lesson that you couldn't have told us in ten minutes on the blackboard?' I gave him the usual answer about learning by doing, but when I had time to reflect I thought 'you know, he's right!' Those boys had not been doing any real science and they knew it. Somehow I had to find a better way of teaching them science.

I went into my prep room and noted down all the most sophisticated equipment I had, even though there was only one of everything: signal generator, oscilloscope, scalar, photo-diodes, neon strobe and so on. It was still important to try to teach the boys common physics knowledge about coefficient of friction, the induction coil, waves, kinematics, etc., so I wrote out a set of workcards which began with the ordinary textbook information in a sentence or two. There was no attempt to deduce the information – it was given tersely as fact. The bulk of each workcard, however, was a set of rather loose questions, for example

- is there any relationship between the length of the spark gap and the input voltage? What has this to do with the design of a motor car?
- does the sensitivity of a photo-cell depend on the colour of the light which illuminates it?
- using the signal generator and speaker to generate sound, and the microphone and oscilloscope to detect it, find out whether there is a frequency at which the system is most sensitive.
- investigate the friction between different shoes and various surfaces on which people normally walk.
- how is the upper threshold of hearing distributed amongst the class?

I can't remember the precise details, but all cards had some follow-up questions about practical application of the students' findings. What these findings would be, I had no idea, and this was deliberate.

The students were grouped mostly in pairs, although some slightly larger teams formed. They were invited to select an investigation and to work on it as long as they liked (ten minutes or three weeks) and to modify it in any way they liked, as I had no idea how it would turn out. When a group felt it was time to terminate an investigation, there would be others waiting from which they could select.

The class was transformed! For the next few weeks, I didn't really know what was going on because the boys kept me so busy fetching and carrying bits of equipment that they decided they wanted. When I did get round to glancing at their books, I found them full of polaroid photographs of oscilloscope traces, masses of clear data, classic graphs, discussions – in short, true scientific reports. And this from boys who a few weeks previously didn't appear to know a graph from graffiti. There was absolutely no content structure for the rest of the year, although C.S.E. was looming ahead. For me, there wasn't much work. The lab assistant had made up kits at the beginning, and these were simply put out with the cards before each lesson. An exciting time for me and if they did badly in C.S.E. it didn't matter very much since they were clearly learning so much in the way of real science, and no one expected them to do well anyway.

They didn't do badly in C.S.E.! Well over half of the class got Grade 1. Several went on to A level Physics, where a few months previously everyone, themselves included, had expected them to leave school for some unskilled job. From Dan I had learned that science had to make sense, that these young men deserved experience with the equipment that real scientists use and that they should be given the opportunity to use it in equivalent ways.

This illustrates in a graphic way the effectiveness of the investigational approach, which many who have tried it would echo. It becomes lear that the cause of students' previous low level of performance in cience is not that they were of low ability and that the work had been o difficult for them, on the contrary their natural scientific potential ad been held back by a highly structured, trivialised form of science hich did not engage with their strengths and imagination.

Ingle and Jennings, in urging that every student should have the xperience of performing genuinely open-ended investigations share ith us the belief that in so doing students will 'gain a better and ore rounded view of the nature of scientific enquiry and of the elation of science to technology than they would obtain from a structured course which limits investigation to a carefully controlled and ven contrived heurism'.[14]

---

xamples of Investigations

---

low do shampoos affect the strength and the setting of hair?
Vhat type of lettering is most legible on an outdoor sign?
low could your kitchen design be improved ergonomically?
low can you get the maximum yield of bean sprouts from mung beans?
Vhat factors affect the loss of water from plants (or the uptake of nutrients)?
nvestigate how reaction time varies with different stimuli and conditions i.e. after drinking tea, time of day).
tudy the distribution and habits of minibeasts in the school area.
nvestigate the environmental factors that affect the heart rate of daphnia.
Vhat causes weathering of limestone (or rusting)? How can it be reduced?
nvestigate the constituents and cooking conditions required to make 'good' ustard.
nvestigate local stream pollution, or air pollution.
ompare the effectiveness of different brands of cleansing liquid.
nalyse the calorific values of different foods and the factors which affect their ate of burning.
nvestigate the composition, structure and strength of local soils.
Vhich is the best detergent/washing powder? What is the optimum concentraion?
tudy the factors that affect the strength of concrete.
low efficient, as an energy converter, is a bow and arrow/electric motor/plant?
Vhat factors affect the control and rate of descent of a parachute?
nvestigate the strength of conkers and how that is affected by differences in size nd age.
Vhat material is most suitable for a shoe sole/a kitchen floor tile/a bicycle brake?
nvestigate the way the shape of card affects its strength.
lake a device for enabling an egg to fall five metres as fast as possible without reaking.
nvestigate the way that the aperture size of a pinhole camera affects the harpness of the image.
nvestigate the ways of harnessing wind or water power.

Before leaving this section on investigation, it would be appropriat to ask what a teacher needs in order to start doing investigationa practical work. We believe very little is needed by way of materia resources and that the most effective and creative investigations ar often those done with the simplest and cheapest apparatus. All teacher needs is a clear view of what investigations are all about, an to be 'switched on' to it. For work in the early secondary years, give imagination and perseverance, a box of very simple tools and a bo containing various kinds of scrap materials (plastics, wood, card etc.) will go a long way. A list of possible suggestions is often usefu to give confidence in starting, and these are now available,[15] bu teachers find that, once they get under way, there is no shortage o topics for investigations that arise both out of the normal course wor and their, and the pupils', fertile minds.

There is, of course, a long tradition of advocacy for the use o individual investigational practical work in school science. Yet i has remained little more than a 'thin red line' in practice. We hav discussed elsewhere how various factors, including the very strengt and status of traditional science in the school curriculum, hav prevented a greater uptake.[16] The current debate over the suitabilit of many science courses will provide the opportunity to rethink an reshape them to include much more investigational work. We believ that the time is now ripe for a much wider acceptance of investiga tions in school science, not just as peripheral extras but as fundamen tal to the whole nature of practical work in science.

Experiences

Experiences are aimed at enabling the students to get 'a feel' for th phenomena being studied. Experiences are often very short, quick exploratory experiments and in consequence their importance i often underrated. But a few minutes experiencing a certai phenomenon, allowing time to think about and discuss it, is ofte time most profitably spent. Moving a finger around and above lighted candle, carefully and thoughtfully, gives a real experience and understanding, of convection currents. Feeling how one's puls rate varies after physical exercise gives a good insight into respiratio recovery. Watching a tiny piece of sodium dance around the surfac of water can lead quickly into most constructive discussion o exothermic chemical changes and reactivity series. The simpl experience, with opportunity to contemplate, to assimilate and t discuss, is invaluable.

Getting a feel for the basic phenomena leads to comprehension and belief. Seeing a crystal growing on a microscope slide gives considerable personal credibility to the theories underlying its regular atomic structure. Watching a tiny oil drop spread out widely, but finitely, causes the student to believe that oil is made up of many tiny molecules, of finite size – unless the elegance of the experiment is submerged in the plethora of incomprehensible mathematics.

It is possible to measure the thermal conductivity of brass (or is it wood or felt?!) using the Searle's bar apparatus, but in so doing students gain no feel for conductivity. How much better for them to hold rods of brass and aluminium in a bunsen and feel heat being conducted at different rates. It is possible to study the way in which starch is broken down into glucose by the amylase in saliva with a series of iodine tests on samples of different 'age' on a dip-tile, but in so doing students can quickly forget what is reacting with what. How much better for them to be given a piece of bread and told to chew it until they can taste it turning sweeter. It is possible to investigate the gas laws with Boyle's or Charles' apparatus, but in so doing it is easy for students to forget that they are studying the properties of air, not mercury. How much simpler for them to feel the increase in pressure as they push in the piston of a plastic syringe or see a length of air in a capillary tube expand as the water in which it is placed is heated. It is possible to study insects through books, or photographs, or mounted collections, but that is nothing compared to the experience of collecting your own in a simple yogurt pot insect trap – and getting a feel for their frequency and distribution as you inspect them with the early morning dew.

Getting a feel for elasticity and fatigue through pulling an elastic band and repeatedly bending the wire in a paper clip, getting a feel for the wave-length of light by looking through scratches in microscope slides coated with aquadag or through the space between two fingers, getting a feel for the smell of ammonia and ethyne, all these are important experiences for a young scientist. It is important too to realise that students will have had a life time of experiencing scientific phenomena outside the school laboratory context and such experiences need to be recalled and utilised in discussion of the underlying concepts.

Many experiments with sophisticated apparatus and quantitative analysis are designed to give a better understanding of certain scientific phenomena. Yet, very often that is hidden from the student, who would appreciate it far more with a simpler experience, obtained at the qualitative, or quasi-quantitative level. Most of the basic laws of nature are elegantly simple (though not always easy to comprehend). Newton's laws of motion, the principles of moments, th conservation of momentum are *not* complicated principles. Yet ver often we make them appear difficult by trying to unravel them from series of secondary data derived from trolleys and ticker timers ruler, weights and fulcrums, or multishot photographs and electronic timers. Once the student has gained a feel for the phenomen of force, acceleration, leverage, etc., through appropriately selecte experiences, the relating principles can be derived simply an clearly through directed thought and discussion around a demonstration.

Much of the basic theory in science is not difficult, but elegantl simple. It is a sad fact that we have often made it appear unnecessarily complicated because of the way we have taught it. We hav hidden the underlying principles in a smoke screen of distractin apparatus and detailed quantitative experimental data. Our plea fo more emphasis to be placed on practical experiences will help t prevent this. Some examples of experiences are shown opposite.

Notes and References

1 J. Ravetz *Scientific Knowledge and its Social Problems*, Oxford Universit Press (New York) 1971 page 79.

2 E. M. Rogers (editor) *Revised Nuffield Physics General Introductio* Longman (London) 1977 page 3.

3 Oxford Educational Research Group Technology Project *Science, Technolog and Society for 11–13 year olds*, mimeo (1983) page 41.

4 P. B. Medawar *Induction and Intuition in Scientific Thought*, Methue (London) 1969 page 57.

5 Assessment of Performance Unit (A.P.U.) *Science at Age 13*, Department o Education and Science (London) 1984 page 4.

6 Nuffield Chemistry *Handbook for Teachers*, Longmans (London) 1967 page 3 and 4.

7 J. L. Moore and F. H. Thomas 'Computer simulation of experiments' *Schoo Science Review*, Volume 64 (1983) pages 641–655.

8 For example, A.P.U., see note 5 above, page 31.

9 CHEM Study *Chemistry – an Experimental Science*, W. H. Freeman (Sa Francisco) 1963 page 1.

10 The Association for Science Education has published a series of pamphlet under the general title of *Primary Science*. They include excellent reviews o work done in schools.

11 R. T. Allsop and B. E. Woolnough 'A Technological Flavour' *Times Educational Supplement* (18.9.1981) page 34.

12 B. E. Woolnough 'Controlled high speed photography of milk splashes *School Science Review*, Volume 51 (1970) pages 891–895.

13 J. Reay, Personal communication through mimeo paper.

14 R. B. Ingle and A. Jennings *Science in Schools: Which Way Now?*, N.F.E.R (Windsor) 1981 pages 81 and 82.

Examples of Experiences

Studying and dissecting plants, flowers and fruits.
Handling animals, vertebrate and invertebrate.
Watching simple organisms (e.g. amoebae) move.
Exploring sense perceptors on the tongue, limits of hearing, aspects of vision.
Testing physiological responses to exercise.
Growing plants.

Growing crystals.
Studying chemical changes with
- colour changes
- precipitate formation
- gas evolution
- energy changes

Synthesising a new compound
- a nylon thread, a synthetic rubber

Getting pure substances from mixtures
- distillation of crude oil

Pulling
- a rubber band
- some copper wire
- a strip of polythene.

Watching Brownian movement in smoke cells and the formation of oil films on water.
Observing interference and diffraction patterns through slits.
Compressing air in a syringe.
Vibrating varying lengths of ruler on the edge of a bench.
Holding a brick above and then in some water.
Forming an image with a hand lens or a pinhole camera.
Pushing door closed from handle side and from hinge side of door.
Moving arms in and out on a revolving stool.
Feeling air temperature above and round a lighted candle.

15 Starting points might include:
(a) Nuffield Advanced Physics *Teachers' Handbook*.
(b) D. R. Browning *Projects in Science and Technology*, N.C.S.T., Trent Polytechnic.
(c) D. Kincaid and P. Coles *Science in a Topic* series Hulton (London) 1973 onwards.
(d) British Association for the Advancement of Science *Ideas for Egg Races*.
(e) *Starting Science*, Central TV.
(f) R. Hawkey *Sport Science*, Hodder and Stoughton.

16 R. T. Allsop, M. Nash, B. E. Woolnough 'Factors affecting the uptake of technological inputs to science education for 11–13 year olds', *Proceedings of Second International Symposium on World Trends in Science Education*, Trent Polytechnic (Nottingham) 1984.

5

Extending the Framework

Hybrid Practicals

We have considered our three aims for practical work and we have suggested a particular type of practical work to match each aim. The question now arises as to whether it is possible for some practicals to be hybrid, with more than one aim being satisfied simultaneously. Our own experience leads us to believe that this may be possible, especially as the student gets more experienced in the laboratory work, but urge caution before use of practical work with mixed aims. In the past great disservice has been done to practical work in trying to use it to teach scientific process and scientific concepts simultaneously. Of course, if practicals are not to be used to try to discover theoretical concepts, there is less danger of this. What is important, however, in teaching is that we should all the time have before us the question 'What is the prime aim?' In other words, what am I trying to do in this experiment? To teach skill and techniques? To give experience of working like a problem-solving scientist? Or to get 'a feel' for phenomena? We need to be careful to ensure that in trying to satisfy more than one aim at a time we do not frustrate the achievement of any one of them.

The student needs to be as aware of the aims of the practical as the teacher is. The student needs to be clear not only about the specific objective of a practical, but also about the over-riding aim, for that will need to be stated so that the student may know how to approach the task. If our aim is to develop the scientific skill of accurate measurement, this will require precision of quite different order from that of a practical aimed at getting a feel for a phenomenon, or one used as a quick test experiment as part of an investigation.

Field Studies

Field studies or 'field teaching' as some prefer,[1] has traditionally received strong support in this country. It has usually been focussed directly on study of the natural environment but the notion can easily be stretched to include studies of museums or industrial locations.

As with work in school science, the first need is for a clarification of aims and objectives. These have not always been fully articulated for field studies, but listings generally include aims such as encouraging a spirit of enquiry, encouraging student initiative and active participation, and the development of hypotheses.[2] Shepley has given three objectives for field teaching in an introductory environmental science course:

1. To provide first-hand investigational experience of the environment both in its capacity of interrelationships and its variety.
2. To illustrate how generally applicable concepts can be derived from observation and investigation of the environment.
3. To provide experience of a range of techniques and instrumentation which can be used in satisfying the first two objectives.[3]

The first and third of these objectives fit very closely with the aims we have stated for practical work. Field studies provide opportunities for developing practical skills and techniques, being a problem-solving scientist through investigation, and having experiences which enable a feel for the various phenomena to be acquired. The reservations we have expressed over the linkage between practical activity and the development of theory hold just as strongly with respect to objective 2 in the listing above. 'Observation and investigation of the environment' is exactly right, 'derivation of generally applicable concepts' will be better carried forward by other methods.

A word should be added concerning the potential of the earth sciences for realising practical work objectives. This whole area of scientific endeavour has been seriously neglected in this country, as compared with, say, the United States and Australia. Practical work in the field in the earth sciences provides opportunities for the full range of aims and types of practical work to be developed, working in an area which appears to be naturally of great interest to students, contrasting sharply with the sterility of much laboratory work in school geology courses. The additional dimension is, of course, time – geological time. The accumulation of evidence from rock exposure in a quarry or a cliff-face in order to interpret past environments, provides a distinctive type of investigation which provides great interest for students. Motivation is frequently mentioned as one of the great benefits of working through field studies – a cautionary note should be added to the effect that this is only likely to be strong where investigational work is involved.

Demonstrations

The use of demonstrations in teaching science has been described as a dying art. Kerr found that the great emphasis laid on small-group practical work had led to a neglect of demonstration, especially with older classes,[4] and this has been accentuated by the emphasis on more individual pupil work in the Nuffield courses even though specific demonstrations were encouraged. In the Ordinary level Biology course, for instance, we find 'the evidence available suggests that, for conveying information and describing concepts and techniques, a demonstration is as efficient as, if not better than, class work. Furthermore it is much quicker and, if used judiciously, can gain more time in the course for more class practical work . . . There is a place in biology teaching for both demonstration and class work. They are complementary to each other.'[5] We would support this argument, indeed would put it less apologetically, for we believe that demonstrations have real value in doing those things which we have argued small-group practical work cannot do. There is an important place for both, fulfilling different and distinct purposes. Garrett and Roberts have provided a comprehensive review of research comparing the relative merits of demonstration and small-group practical work and have found, not surprisingly, no conclusive result, as both teachers and researchers are unclear of the specific, and different, appropriate aims for the two.[6] Pre-eminently, demonstrations provide the ideal opportunity for the teacher to link the practical reality with the underlying theory. By closely and carefully directing the argument around the demonstration the teacher can indicate and illustrate the abstract principles involved. The important, relevant aspects can be emphasised and the secondary irrelevancies dismissed. In this way the theory and the practice can be effectively tied together. Such demonstrations may be used to introduce, to develop or to recap on a series of experiments, but in each case, because the teacher is clearly in control, the relevant part of the argument can be stressed. The ancillary purposes of demonstrations are more obvious: the ability to show experiments that are too dangerous or expensive for the students themselves and the dramatic and indelible impression left in the memory by a really spectacular demonstration. We suspect that for many practising scientists and science teachers the impression left by one or more dramatic demonstrations experienced in their formative youth was highly influential in determining their subsequent career in science.

So we would see demonstrations not as being an alternative to small-group practical work, but as having a vitally important, complementary role, helping the students to make the vital links between reality and abstract theories, and enabling the student to build up more structured and interconnected cognitive frameworks.

Use of Laboratory Apparatus and Time

We have been critical of some types of practical work where quite sophisticated apparatus has been developed to try to discover a piece of theory. We have suggested that the theoretical concepts have often been obscured by the clutter of complex apparatus, by the manipulative skills required to use it and by the emphasis on collecting precise quantitative data. We have suggested that simpler apparatus would enable pupils to acquire a better feel for phenomena. However, this does not mean that the more sophisticated apparatus should not be used; on the contrary, it can provide excellent facilities for investigational projects. The physics apparatus, for instance, originally destined to discover the basic law of motion – ticker timers, multishot photography, milli-second timers – provides marvellous tools for investigating real problems. Ticker timers can be used to analyse accelerations, and thence the forces, associated with athletes' sprint starts or weight lifters' lifts. Multishot photography can be used to analyse weight lifters' action, or the way that a splash is formed or a piece of glass broken. Milli-second timers, and trolleys – even linear air tracks – can be used to measure the speed of an air gun pellet and to see how it varies with distance. The range of investigations appears endless once these tools are separated from their role of discovering theory. The use of pH meters and colorimeters for investigations has already been very well promoted through the environmental testing probes, though simpler laboratory apparatus could be used for such investigations.

Indeed it should be stressed that, although sophisticated apparatus is sometimes useful in investigational projects, many, perhaps most and probably the best, of the investigations are done with the very simplest apparatus. For the simpler the equipment used, the more flexible it may be. The simplicity will not detract from the essence of the practical. A simpler force measurer can easily be made from a piece of dowel, a piece of plastic tube and an elastic band and then used in an investigational project with different loads being pulled on different surfaces. It will first have to be calibrated, a valid activity itself. The student will need to consider the range within which it will give reliable measurements. When the elastic band breaks, it can easily be replaced. On the way, a student will have achieved a real feel for the phenomenon of force, with a much better understanding than that achieved by taking a smartly cased newtonmeter from a drawer. There is here an important element of demystification, which is often necessary after our students' exposure to endless 'black boxes'. There is virtue in using, say, plastic yogurt pots instead of glass beakers or copper calorimeters. Apart from being cheaper, they are more flexible and mobile and may be easily modified, and reshaped. They also help to remove some of the air of mystery surrounding science and help students to relate it more to their everyday world. Much of the best and most creative of science work is done with simple, 'string and sealing wax' apparatus. Perhaps we should be saying that the well-equipped lab is not the one with a lot of sophisticated hardware, but the one containing boxes of scrap material and simple tools. The other 'cheap' source of 'scientific apparatus' is, of course, the students' natural environment – the real world. We often complain of the problems of transferring school laboratory science into the real world; if we started our science more in the real world, using the students' experiences and their environment as our laboratory, we would have fewer problems here. We should seek to make more use of the home, and the local environment, both as the test-bed and the illustration of our science. Not only is it cheaper, it is also more versatile and effective for promoting learning.

One of the implications of our argument is a shift in the way that time is spent on practicals. The exercises we suggest are often short, many of the experiences would also only take a small part of a lesson. Investigations would usually take longer, frequently more than a double lesson and often a series of lessons in sequence. The extra time for these could readily be found by reducing the time currently spent on the ineffective 'cookery book' type of practicals being used in the vain hope that they will reveal the theory.

The Place of Technology in Science

There have been many recent exhortations for education in general, and science teaching in particular, to become more technological. Not all have been clear about what they mean by technology. We see technology as having two quite distinct elements, one of relevance to science practical work, the other not. The first and central element in technology is problem-solving, the open-ended design process in which a problem is established and analysed, possible solutions considered, and the optimum solution selected, implemented and evaluated. This is defined most commonly as 'a disciplined process using scientific, material and human resources to achieve human purpose'. This is very similar to the investigational approach of a problem-solving scientist and matches very closely our approach through investigations. Some distinction has been made between technological investigations which are set in the context of some specific human need and science investigations which may be followed purely to satisfy curiosity. Technological investigations are more likely to involve making some device than are scientific investigations, which may well use standard laboratory apparatus, but this distinction is not an important one. Students can experience and practise involvement with the process of technology in a variety of ways in school, through technology and hobbies clubs, through science fairs or industry-sponsored competitions, through specific technology courses, or through Craft–Design–Technology or Science lessons becoming 'technologically flavoured'. We are concerned here with the ways in which science teachers can develop it through investigations in science lessons.

Fig. 5.1 The process of technology.

Source: **A.R. Marshall (editor) *School Technology in Action,* English Universities Press (London) 1974 page 6.**

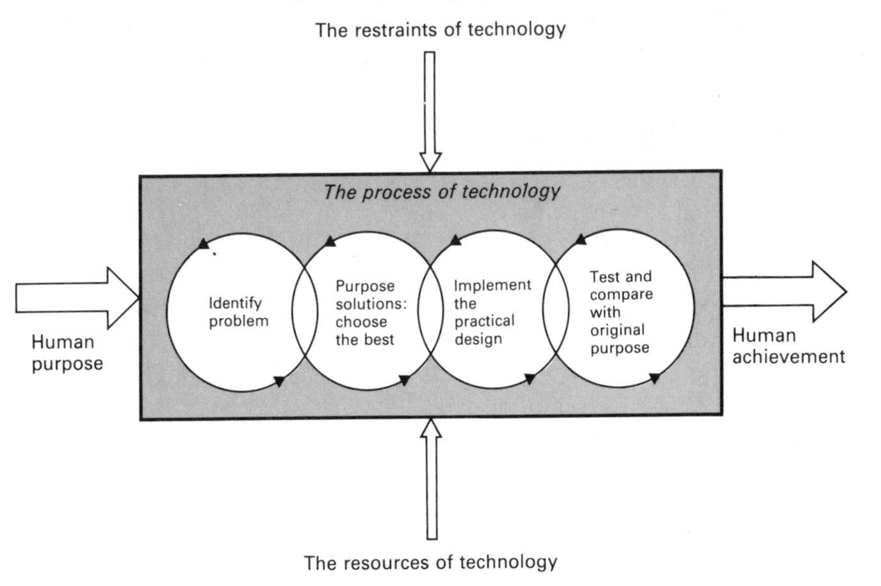

The second element of technology relates to the application of science and the implications of technological processes and products for society. This field is vitally important for science teachers to be involved with, but does not normally include practical work and will not be considered further here.

There is an alternative approach to technology being introduced to schools which we believe is quite inappropriate and divisive. It is essentially a vocational approach, where sponsors argue that as we are living in an increasingly technological world, requiring an increasingly technologically skilled labour force, schools should direct themselves at giving students a training in a range of technological skills. Hence a list of technological skills is drawn up, linked to the perceived job requirements of particular forms of employment, with the aim that students may be prepared for those jobs. We believe that schools are not the right place for such specific vocational training and that past and current performance illustrates how schools cannot keep pace with either the needs or the techniques of a rapidly changing and diverse employment scene. We believe that schools *are* appropriate institutions for providing a broad education for all students, and our arguments for encouraging and developing the skills of a problem-solving scientist, or a technologist, should be seen in that light. Incidentally, we believe that this also makes the most appropriate and constructive base for young people to take with them into their particular form of employment, in which they can most appropriately acquire the specific job-related skill required.

Science in Primary Schools

One of the most encouraging developments in science education over the past few years has been the growth of scientific activity in the primary school (age 5–11). Both Nuffield Primary Science and Schools' Council Science 5–13 grew out of the child-centred, integrated approach of the primary schools. They focussed on the child's area of interest and developed an investigational approach to aspects of the environment or topic under investigation. Though the formal uptake of these two schemes has been small, largely due to lack of confidence and experience by the teaching force in primary schools and the diffuse nature of the curriculum materials provided, they represent a most important and significant approach. At the centre of this approach is the belief that the child can, and should be encouraged to work as a problem-solving scientist, as we have argued

Fig. 5.2 The aims of Science 5–13.

Source: **Schools Council Science 5–13 *With Objectives in Mind*, Macdonald Educational (London) 1972 page 59.**

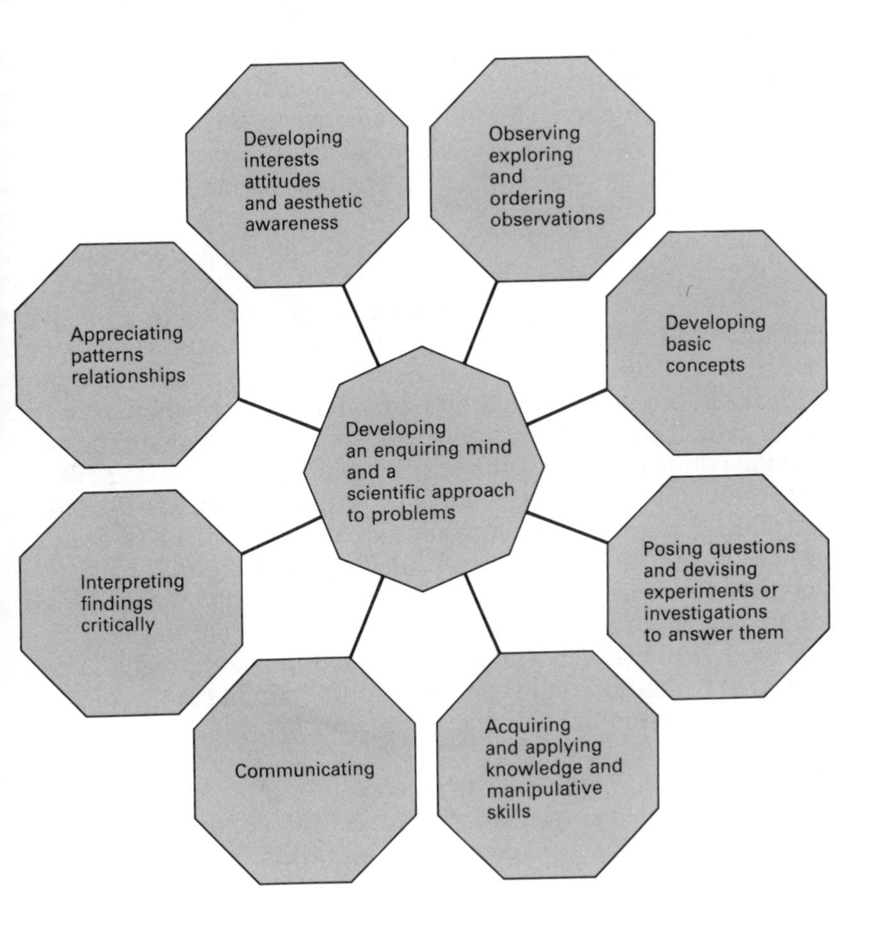

earlier. It concentrates on the processes of science, and leaves the content as of secondary importance and the concepts as too difficult and inappropriate. Science 5–13 spells out the main aim: 'developing an enquiring mind and a scientific approach to problems'.

The Assessment of Performance Unit describes their view of primary science as a rational way of finding out about the world, involving the development of a willingness and ability to seek and use evidence; the gradual building of a framework of concepts which helps to make sense of experience; and the fostering of skills and attitudes necessary for investigation and experimentation.

Her Majesty's Inspectors of Schools, the Association for Science Education and the British Association of Young Scientists have all done much to encourage this investigational approach to science teaching, which develops the habits of a problem-solving scientist and builds up the tacit knowledge in appropriate contexts which will form the foundations for future understanding and problem-solving. As the children investigate their own environment, whether in the school field, the classroom, or the neighbouring building site, and ask questions of it, as they test the properties of materials that are relevant to their topic of clothing or housing, as they devise experiments to study flotation, water resistance and sail design when they are studying the topic of boats and travel, they are learning to act like real scientists. The science in this approach can be fitted naturally into the broader topic that the class is studying and does not become a separated, artificial activity.[7] Clearly this type of approach is very much that which we would want to encourage throughout a scientific education. It exemplifies our exercises, investigations and experiences.

There is another approach to science in primary schools which might be described as 'secondary science watered down'. It centres on scientific knowledge and concepts, and aims to impart this to the children. Inevitably it leads to worksheet-directed activities or didactic teaching from the teacher and such an approach does not naturally develop from the children's own interests. We believe that such an emphasis on content is inappropriate, and any hope of the children appreciating the abstract concepts is naively unrealistic. It also misses the opportunity of encouraging the children's natural tendency of being 'pupil as scientist', and building upon their tacit knowledge.

As science becomes stronger in the primary school we see afresh the perennial battle concerning the nature of science and science teaching that has continued throughout the hundred years or so of its practice. At the time when we are encouraging more of the investigational approach to science to move up through the secondary school, others are trying to move the content-based, impersonal approach to science down into the primary school. The struggle between science the useful and science the beautiful needs to be resolved in the primary school too!

Notes and References

1 A. V. Shepley 'Towards more effective field teaching' *School Science Review*, Volume 55 (1974) pages 817–822.
2 Report of Study Group on Education and Field Biology *Science Out of Doors*, Longman (Harlow) 1963 page 22.
3 In Shepley, see note 1 as above, page 819.
4 J. Kerr *Practical Work in School Science*, Leicester University Press (Leicester) 1963.
5 Nuffield Ordinary Level Biology *Teachers' Guide*, Longman/Penguin (London) 1966 page xiii.
6 R. M. Garrett and I. F. Roberts 'Demonstration versus small group practical work in science education' *Studies in Science Education*, Volume 9 (1982) pages 109–145.
7 See D. Kincaid and P. Coles *Science in a Topic*, Hulton (London) 1973 onwards.

6

Implications for Science Teaching

What then are the implications for science teaching in schools and colleges?

The Process and Content of Science

Fundamentally we must recognise that science teaching is concerned with both the content of science *and* the process of science; with what Layton refers to as 'an understanding of the mature concepts and theories of science and an understanding of the processes by which scientific knowledge grows'.[1] Both are vital for a full scientific education. It is important that students acquire the habits and skills of a practising scientist, and learn to approach a problem 'scientifically'. It is also important that students build up a personal mastery of the knowledge and theories of science that form the body of our scientific culture. The approaches needed for students to acquire these two aspects of science are not the same, and the common practice of trying to master both simultaneously can lead to inadequate understanding of each. It would be possible to acquire a knowledge (albeit shallow and limited) of much scientific information without doing any practical work at all. It would not be possible to learn how to work as a problem-solving scientist in such a way. So, if we are concerned to teach the processes of science, practical work is vital. But it is also vital that students, at some stage, acquire a meaningful understanding of the concepts and content of science and for this we may need to develop alternative strategies more fully, and recognise that a mastery of abstract concepts needs hard mental exercise and discussion at the abstract level. We have argued that too early an abstract formalisation of scientific concepts can produce premature, stunted growth that hinders rather than helps fuller assimilation when the student is more intellectually mature. We have also argued that when a fuller degree of cognitive maturity has been acquired, the student should not always be restricted in developing mastery of the abstraction by concrete experimentation. The question arises as to what emphasis should be put on which aspect at the different stages of a science education. We welcome the introduction of science teaching in the primary schools, but reject the idea that this should be an opportunity to teach formal science content or concepts. Even the statements of such aims as 'laying foundations for the development of scientific concepts later' encompass the danger that we will again have a content-led curriculum likely to introduce concepts prematurely. We would argue that the early scientific education of students should be concerned not with the theoretical content but with developing processes of science. This should concentrate on investigational practical work centred on the student's own environment. The context in which such work would be developed would need to be ordered but would be determined by the local environment, the student's and the teacher's interests, and those themes being studied in the curriculum as a whole. Hence the scientific way of looking at things would be seen as a vital and integral part of the whole curriculum. The science teacher's aim here would be to develop in the children the habit of working like a problem-solving scientist, and to enable them to gain a wide range of tacit knowledge, and confidence in using it, in areas of scientific interest. It is the time to open their eyes to the world around them and to encourage them to explore and investigate it scientifically. We believe that such investigational, environmental science should form the basis for science teaching up to the age of 13 years. Subsequently, having developed a wide tacit knowledge of various, carefully chosen areas of the environment, and acquired skills and habits of enquiring about the world scientifically, the students should be led to focus more fully on the concepts and the content of science. Much of this focussing will be done, unashamedly, in a non-practical context – through discussions, debate and application. These will be of increasing importance to those likely to take their science further, either vocationally or culturally. For many, however, these concepts will never be of central importance, of much more value will be the ability to work as a problem-solving scientist using tacit knowledge directly, and for these especially science education should concentrate on the processes rather than the decimated content of a watered-down academic course. Investigational work, however, should not be excluded from the curriculum of the academically able from the age of 13. On the contrary it should retain its central position throughout as fundamental to a maturing scientific education, both as an end in itself and as an aid to the students making sense of the theoretical concepts. A student, whether at school or college, who only develops cognitive head knowledge, will only be developing half of a science education.

Personal and Tacit Knowledge

The second implication of our discussion relates to the vital need for students to 'personalise' the knowledge for themselves. Scientific knowledge should no longer be seen as something external which can be added on to a student's cognitive framework. It must be personalised, be worked on individually, if it is to acquire meaning. Students cannot receive knowledge and understanding passively – they need to be active in the learning process. Barnes used the phrase 'talking oneself into understanding',[2] this can be extended into 'writing oneself into understanding', or more generally, 'working oneself into understanding'. Such 'working' may come through practical experimentation or it may come through discussion, debate and theoretical exercise. As Bell and Driver put it, 'learning science is not a matter of passively absorbing information but of actively constructing for oneself an understanding or meaning of the situations presented'.[3]

This leads us into the vital importance of tacit knowledge, both as a means of direct problem solving and also as an essential foundation for a genuine understading of theoretical concepts. It is part of the cultural tradition in Britain to give high regard to 'knowledge for its own sake', to those who *know* a lot and to give much less regard to functional knowledge and to those who *can do* a lot – science the beautiful as distinct from science the useful, the intellectual as distinct from the practitioner, the scientist as distinct from the engineer. Much practice in schools, with the restrictive influence of a university-dominated examination system, reinforces this unhealthy dichotomy. We want to re-emphasise the value of tacit knowledge, a personalised knowledge that comes through experience and which is of fundamental importance. We often pretend that focal, articulated knowledge is the highest aspiration and that having acquired this we use it to solve problems. In reality most problem solving is done directly through tacit knowledge, acquired through personal experience. Similarly an explicit, abstracted, articulated, theoretical understanding of the fundamental concepts of science can only have real meaning if that too is rooted and grounded in tacit knowledge. As Polanyi says, 'all knowledge is either tacit or rooted in tacit knowledge. A wholly explicit knowledge is unthinkable'.[4] So we would wish to give primacy to tacit knowledge, and ensure that the practical experiences are so directed to a broad range of appropriate contexts that students acquire a wide range of tacit knowledge which will have both direct and formative value.

Relationship between Practical and Theory

We have spoken earlier of the dangers of muddling together practical and theory so that the aims of each are not distinguished. In this way, we have argued, practical work has become distorted, and theoretical understanding not satisfactorily achieved. Our view is that, initially, each should be considered separately from the other so that we might establish a self-sufficient rationale and *modus operandi* for each. However, there will be an important interaction between the two, between the experiences gained by the student in doing practical work and the theoretical understanding achieved of the underlying concepts. It is this area of interaction that we will now examine, to see how practical experiences may help theoretical understanding, and how theoretical understanding makes practical work more effective.

Our starting point is with the experiences that the student has had, both through practical work in the laboratory and also through the day-to-day involvement with the physical world which has been a formative part of the student's life and thinking since birth. Much of school science teaching has ignored this area, but it is evident that such experience has a very important influence on the development of the student's thinking; indeed the sorting out of a life time of experiences and developing theoretical 'models' which enables sense to be made of the world has been proceeding for many years before entry to a science laboratory. The child has been active as a scientist – observing, testing, hypothesising and evaluating – all its life and comes to a science laboratory with a mind full of relevant experience and insights which may, if ignored, prove very difficult to alter. The idea that a student commences a school science course with a mind empty of any scientific knowledge and understanding is false, and educationally disastrous.

So the student arrives at the science lesson with a range of scientific experiences, which have already given a wealth of tacit knowledge and also various insights. Such insights will be partial, often ill-formed, but personally meaningful. We could call them cognitive 'bits and pieces', ('baps'), as they do not form a consistent picture or a whole framework. Cognitive 'baps' are small, localised, independent insights which may even be mutually contradictory. It is a science teacher's job to help the student to build from these a larger, logical, consistent framework which will more satisfactorily explain a wide range of phenomena in the scientific world. 'Teaching needs', according to Osborne and Wittrock, 'to take fully into account pupils' perceptions and viewpoints and, where appropriate, to attempt to modify or build on, but certainly not ignore, children's ideas.'[5]

The first step in this process is to help the students to articulate, and to develop through discussions, those insights which are already there. The teacher needs to draw out, and clarify, those insights which are appropriate, and direct the student's thinking along the right lines. Thus the students will be able to come to practical work with 'meaningful insight' which will enable them to select, and make sense of, the mass of information that will be received when doing an experiment.

Subsequently, hard discussion, illustration and application will be needed, away from the apparatus, to enable the student to develop and refine those insights into a more mature, and useful cognitive framework. The concrete reality of the laboratory is not, now, helpful in developing and applying abstract concepts into new theoretical problems. In speaking of the generative learning process, Osborne and Wittrock say that 'when we give information to pupils, or answer a pupil's question, our statement or explanation may help a pupil but it can only help, or lead to a new perception, when the pupil does something with the information. That something is generation, the act of relating the pupil's knowledge, logic and experience to parts of the statement or explanation, and the construction of meaning'.[6] It is in both discussion and application, either to theoretical or practical problems, that the student is able to 'do something' with the information and extend his understanding meaningfully.

The students, having been enabled to develop and refine their understanding through discussion, writing and application, will now be able to take it into new practical investigations and use it in suggesting and interpreting more perceptive solutions. Such investigations and experiences will more fully personalise understanding in a way which will never be forgotten. So we have suggested that this line of practical–theory interaction is a process of successively taking partially formed insights obtained from past experiences, articulating and refining them through discussion and application, and then taking them back into further practical experience and investigations to personalise and use them as more mature insights.

There is, however, a further line to practical investigations and

Fig. 6.1 The practical–theory interaction.

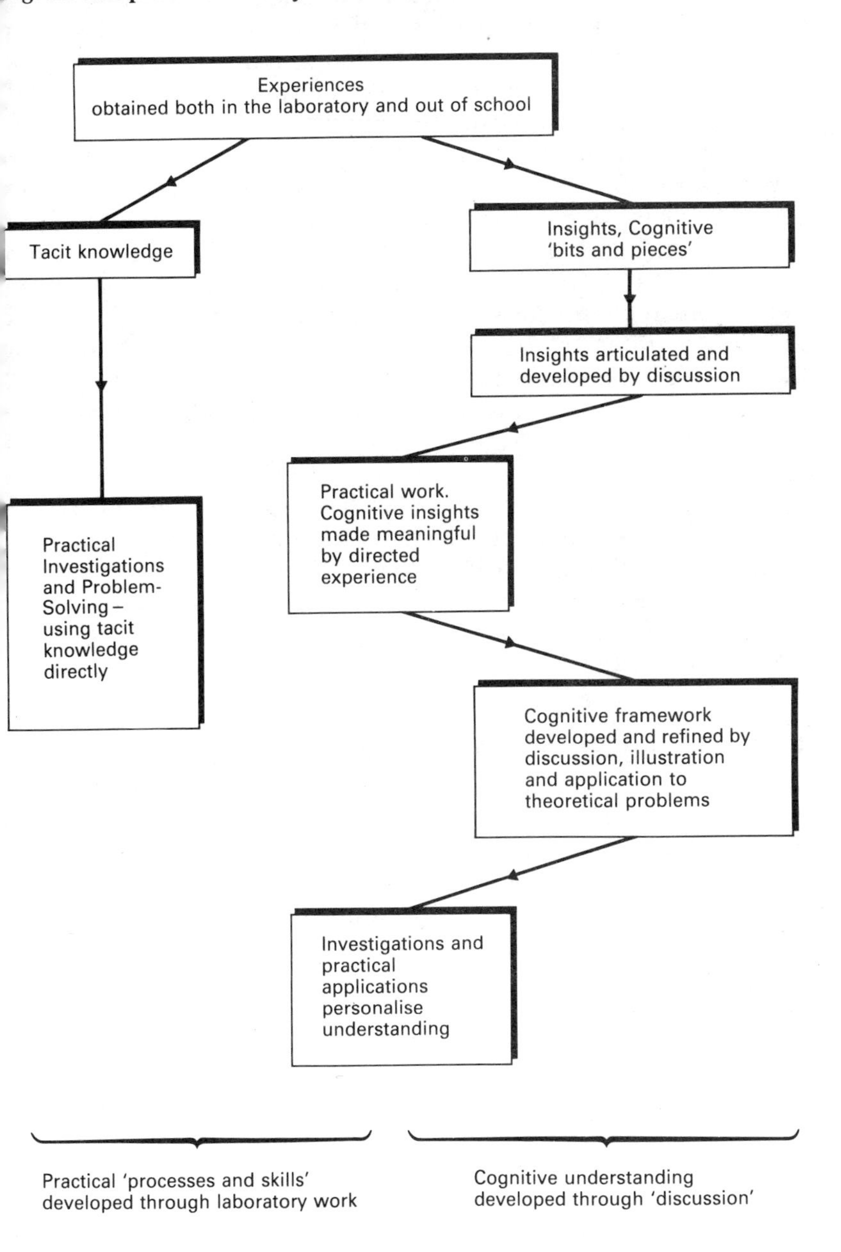

problem solving which does not rely on articulating the underlying theory. Much scientific research and development relies directly on tacit knowledge and the habit of problem solving. We often pretend that science advances by articulating and formalising an understanding of basic theory and then applying this to a new situation. Much of science is a craft activity, where the scientist takes the tacit knowledge acquired through experience and uses it through the habitually acquired process of problem solving. As this is true for real science we should encourage it more for our students in school science too. Students will thus acquire a fund of tacit knowledge that can be applied directly in practical investigations and problem solving both to obtain solutions to the problem attacked and to develop the habit of working like a problem-solving scientist.

We have tried to analyse the relationship of both practical work and discussion to the two aims of enabling the student to acquire the process of science and the mature cognitive understanding of the content and concepts of science. We have summarised this in Fig. 6.1. It should be stressed, however, that although this implies a linear process in practice it will be more cyclical. The experiences and insights gained at the end of the line will all be fed back into the fund of tacit knowledge and cognitive 'bits and pieces' to be applied and developed in further activities.

Means and Endpoints for Science Teaching

We started our discussion with a plea that we should be clearer about, and more discriminating in, our aim for science teaching, and that practical work in particular should be more carefully analysed to assess which aims it could best fulfil. Perhaps we should conclude our discussion with a summary, drawing together the important endpoints for science teaching, and the way that the practical activities relate to them.

We would start with the endpoints. Our goals for science teaching should be that students should acquire both a mature cognitive understanding of scientific knowledge and concepts and also a mastery of the skills of a practising scientist, which would involve both the development of basic laboratory skills and techniques, and also development of the habit of working as a problem-solving scientist. For a student to receive a full education in science, there must be a mature development of both the skills of a scientist and the understanding of its content. There will, of course, be other aims whereby wider educational aims can be achieved *through* science, but for an education *in* science to be true the fundamental nature of science the student must have mastered both the skills of a problem-solving scientist working in a laboratory and also have gained a mature cognitive understanding of a significant part of our cultural heritage of scientific knowledge.

How then do we reach these endpoints? We have discussed this more fully in earlier sections, suffice here to summarise appropriate means and ends through Fig. 6.2. Exercises lead to the development of laboratory skills. Basic skills will be needed in investigations which will use experience gained to develop problem-solving skills. Experiences will also form the foundation for discussion, development and application which leads to cognitive understanding. There will be other links, notably the feedback of cognitive understanding into investigations which will suggest new lines of approach and interpretation, but the simplicity of the framework concentrates the thinking on the main, distinctive threads.

Fig. 6.2 Means and endpoints for science teaching.

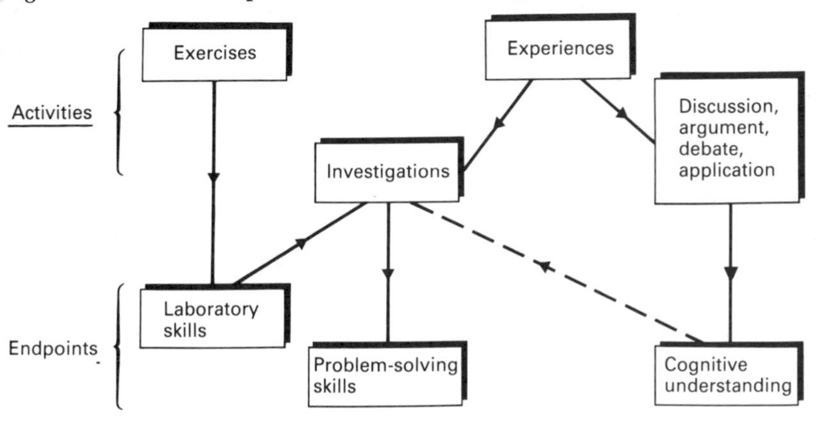

New Aims for Science Teaching

Throughout the discussion in this book we have concentrated our aims on education *in* science. We have done this deliberately, partly to limit the scope of the discussion, and partly because we believe that such an education is an important part of everyone's education. Though the emphasis might vary for different students at different stages, all should acquire the skills of 'useful science' and appreciate something of 'beautiful science'. But others have recently been stressing the value of education *through* science and with this we would agree, on the assumption that the science itself (as we have analysed it) does not get lost. Recently many significant statements have suggested new aims for the teaching of science.[7] In addition to those covered in the previous section, three groups of aims recur: aims relating to an appreciation of the relationship between science and society and how that has developed; aims relating to the acquisition of decision making skills, both at the individual level and as a citizen of our society; and aims relating to the development of the student's personality, to enable a sense of personal identity and dignity to be developed as well as the desired characteristics of independence, co-operativeness, honesty, curiosity, industry and commitment. While recognising that these aims are not exclusively scientific, including aspects and approaches common to other subjects, we would welcome them as entirely appropriate for inclusion in science lessons. We would hope, however, that science teachers would not claim a monopoly of them, but seek to develop them in consultation with other colleagues as part of 'whole curriculum' planning.

How far does practical work have any part to play in fulfilling these new aims? For the first two aims, the answer is rather little. Issues relating to science and society would only be trivialised if these were tied to the practical work that students were able to do, though certain investigations are likely to raise questions which relate to the use, or abuse, of science in society. Though the decision making that will be made in some practical investigations will involve non-scientific factors, such as aesthetic, economic, moral and societal considerations, these will inevitably be handled at a relatively low level. An introduction to the way that decisions are made in our democratic society needs to be tackled more directly – through discussion, simulation, case studies, and games. But those aims relating to the development of a sense of personal identity and pride can have real fulfilment through the type of practical investigation we have advocated. Closed, content-dominated practicals leading to the 'teacher's right answer' will be inhibiting to the development of a student's independence and self-confidence. Investigational practicals based on problems the students can relate to, and the recognition and encouragement of students as scientists needing to personalise their knowledge and seeking to make sense of their world, will go a long way to develop mature and autonomous personalities, sensitive to the resources and the limitations of science.

Alternative Teaching Strategies

In our discussion of the place of practical work in science teaching, and consideration of what practical work can and cannot do, we have inevitably hinted at alternative teaching strategies. Practical work is essential for fulfilling certain fundamental aims of science teaching but is quite inadequate and inappropriate for others. There is not space here to discuss fully the range of such alternative strategies, but we would want to indicate some which we hope might be developed. Many teachers believe they are only teaching science properly if their students are 'doing practical'. Such is far from the truth.

Perhaps the role of language is central to the development of appropriate ways for a student to personally make sense of the world. Head argues that scientific practical work should be done to satisfy 'learning through language' aims, in that by doing small-group practical work students have the opportunity to discuss together the work they are doing and, thereby, to talk themselves into understanding.[8] Undoubtedly this is a very important part in the learning process, so important that it ought not to be left to chance that it might occur during a practical. Other teaching strategies need to be specifically developed in the science lesson whereby discussion, debate, and student talk are deliberately encouraged to further this vital part of learning. Science teachers need to encourage small-group discussion, either with the whole class or a small group, but there is no way that a teacher could, or should, direct all the student discussion that goes on in the class. It needs, perhaps, an act of faith by the teacher that students can, and will, talk themselves into understanding even without [perhaps better without] the teacher's direction. And what has been said of student discussion applies also to writing – students need to be given the opportunity, through free, loosely structured assignments, to write themselves into understanding. The greater use of directed reading (from books, articles, newspapers, even scientific papers), of debate, of case study discussion, of decision-making simulation exercises, of portfolio projects, of visits, of library studies, of interviews, even of drama, will all benefit the students' understanding and appreciation of the work. Science teachers can learn with profit from the range of strategies used by their colleagues in other subjects who cannot fall back on the ubiquitous experiment to keep the student occupied.

Implications for Teacher Training

Clearly such changes as we have suggested will not occur unless the science-teaching profession is convinced, and this will necessitate the appropriate training of teachers at the initial and, even more, at the post-experience stage. Most science teachers have themselves been brought up on a diet of content-dominated, cookery book type practical work, and many have got in the habit of propagating it themselves. There is still too little evidence of investigational, problem-solving, science in schools. Teachers need to change both their perspective of the nature of scientific activity and, then, their practice in the schools. If this is to be done they will need first to be dissatisfied with current practice and secondly to be convinced of the value of an alternative approach. We believe that, as teachers ask themselves about the aims they hold for their practical work, and further question how effective their current practice is, they will begin to be more selective. The growing evidence about the levels of understanding and skills attained by students studying science should also cause teachers to seek ways of improving their own performance. So the first stage in developing the teaching strategies of science teachers, be they trainee, post-experience or Head of Department, is to provide opportunities for them to study reflectively the insights now available to us about children's attainments in science. The second stage is to be persuaded of the value of a more discriminating, and a more investigational, approach to practical work and this will only be achieved as such practice is experienced in action either in other schools or, better, in their own schools as they 'have a go'. We are convinced that teachers, once having introduced the exercises, investigations and experiences in their practical work with their students will themselves be convinced of their value. We believe that we can give students a better education both *in* and *through* science. We hope that science teachers will find this book both provocative and stimulating, and help them towards providing a more satisfying pattern of practical work for their students.

Notes and References

1 D. Layton *Science for the People*, George Allen and Unwin (London) 1973 page 176.
2 D. Barnes *From Communication to Curriculum*, Penguin (Harmondsworth) 1976 pages 25–31.
3 B. Bell and R. Driver 'The children's learning in science project' *Education in Science*, Number 108 (1984) pages 19–20.

4 M. Polanyi *Knowing and Being*, Routledge and Kegan Paul (London) 1969 page 144.

5 R. J. Osborne and M. C. Wittrock 'Learning Science: a generative process' *Science Education*, Volume 4 (1983) pages 489–508.

6 R. J. Osborne and M. C. Wittrock, see note 5 above.

7 For example,
(a) *Science Education 11–18*, Royal Society 1982
(b) *Science Education in Schools*, Department of Education and Science 1982
(c) *Education through Science*, Association for Science Education 1981
(d) *Science Education 11–16*, Secondary Science Curriculum Review 1983.

8 J. Head 'What can psychology contribute to science education?' *School Science Review*, Volume 63 (1982) pages 631–642.

NAME INDEX

UBJECT INDEX